*Bubbles, voids, and bumps in time:
the new cosmology*

Bubbles, voids, and bumps in time: the new cosmology

EDITED BY
JAMES CORNELL
Publications Manager,
Harvard–Smithsonian Center for Astrophysics

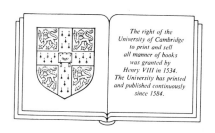

*The right of the
University of Cambridge
to print and sell
all manner of books
was granted by
Henry VIII in 1534.
The University has printed
and published continuously
since 1584.*

CAMBRIDGE UNIVERSITY PRESS
CAMBRIDGE
NEW YORK   NEW ROCHELLE
MELBOURNE   SYDNEY

Published by the Press Syndicate of the University of Cambridge
The Pitt Building, Trumpington Street, Cambridge CB2 1RP
32 East 57th Street, New York, NY 10022, USA
10 Stamford Road, Oakleigh, Melbourne 3166, Australia

First published 1989

Printed in Great Britain at the University Press, Cambridge

British Library Cataloguing in Publication Data

Bubbles, bumps, and voids in time: the
new cosmology.
1. Universe. Origins
I. Cornell, James
523.1'2

Library of Congress Cataloguing in Publication Data

Bubbles, voids, and bumps in time: the new cosmology / edited by
James Cornell.
    p.   cm.
Includes index.
ISBN 0 521 35297 5
1. Cosmology.   2. Astrophysics.   I. Cornell, James.
QB981.B79 1988
523.1 – dc 19   88-15606   CIP

ISBN 0 521 35297 5

PN

PREFACE

At first within the darkness veiled in darkness,
Chaos unknowable, the All lay hid.
Till straitway from the formless void made manifest
By the great power of heat was born the germ.

Rig Veda X.129. Translation by Raymond Van Over, editor,
Sun Songs; New American Library, New York, 1980.

Every culture, from the ancient tale-spinners of the Indus Valley to
the modern technocrats of the Silicon Valley, has held its own unique
view of the cosmos. The mysterious creation of the universe, the
beauty of its apparent perfection, and the terror of its uncertain end
have inspired wonder, awe, and fear. In turn, art, religion, and
science became channels for redirecting the wonder and dissipating
the fear.

Astronomy, the oldest science and offspring of astrology, cannot be
separated completely from more philosophical contemplations about
the cosmos. In fact, despite its observational base, its mathematical
underpinning, and its reliance on supposedly immutable physical
laws, modern astronomy's vision of the cosmos remains colored with
some of the mysticism of the first stargazers.

In part, this is because astronomers still remain at great disadvantage in their search for understanding. Almost all astronomical information is indirect and secondhand, based on the analysis of radiation
arriving at Earth only after an incredibly long journey through a
little-known medium that can distort, disrupt, and change the
messages from the stars. Worse yet, the information received is
woefully incomplete, with most celestial radiation blocked from view
either by the Earth's atmosphere or by the inefficiency of our detectors. Moreover, as will be obvious from reading this volume, perhaps
as much as 90 percent of the material in the universe – whatever it
might be – is simply invisible.

By necessity, then, much of what astronomers claim to know about
the history of the universe is speculative and, according to some
critics, no closer to reality than the epics of early Hindu poet-priests.

Still, recent advances in both ground-based and space-borne telescopes, coupled with improved detectors and enhanced computing capability, have allowed astronomers to look both deeper into space and farther back in time, thus approaching ever closer to the crucial moment of creation. Similarly, new results from particle physics have provided insights into the extraordinary processes that may have shaped the early universe. Thus, while major gaps in knowledge still remain, astronomers think they know at least how – if not exactly when – the universe could have began, and how – if not exactly when and if – it might end.

Such information provides the framework for a distinct field known as cosmology. Once the province of poets and philosophers, the study of the creation, evolution, and possible fate of the universe has, for the past half-century, been a legitimate subject for astrophysicists.

This book presents a comprehensive view of our universe as seen through the work of six leading cosmologists. Be prepared, however, for some challenges – and some disappointments. The subject is obviously the grandest in science and its theories, arguments, and proofs among the most complex. At the same time, much of the business here is unfinished. Few of the questions raised have been answered to everyone's satisfaction – and perhaps will not be in any reader's lifetime. In fact, this volume is best viewed as a 'state-of-the-universe report,' a current summing up, almost certainly subject to revisions.

The concept of this book was born in late 1985 when three astronomers at the Harvard–Smithsonian Center for Astrophysics published the results of a decade-long survey of the redshifts, or recessional velocities, of some 2000 galaxies. Their three-dimensional map of a 6°-wide slice of the sky over the northern hemisphere revealed that the galaxies seemed to be distributed on the surfaces of huge, bubble-like voids, some more than 100 million light years in diameter. This dramatic finding captured the popular imagination, probably because it conjured up images of a soap-bubble universe. More important, the findings raised serious questions about prevailing scientific notions of large-scale structure in the universe.

For more than a decade, the Center and the Boston Museum of Science had cosponsored a series of free public lectures on astronomy supported in part by the Lowell Institute. The new 'slice of the universe' suggested that the theme of modern cosmology might be a natural one for the Spring 1987 series. Indeed, the title of the series (and this volume) attempted to encapsulate some of the major advances in the field. The 'bubbles,' of course, are from the three-dimensional mapping by Margaret Geller and her colleagues. The 'voids' referred to earlier research by Robert Kirshner that first revealed a vast area devoid of galaxies seen in the direction of the Constellation Bootes. More obliquely, perhaps, the voids were suggestive of the pioneering work of Vera Rubin on galactic haloes, which appear to have large quantities of 'invisible' dark matter. Finally, the 'bumps in time' was a popular interpretation of Alan Guth's elegant theory that the very early universe experienced a sudden, brief, and truly gigantic inflationary period. Bracketing these speakers were Alan Lightman, who provided a historical and philosophical context for the modern cosmological views, and James Gunn, who offered a look at the prospects for better understanding to be gained from Space Telescope and other instruments planned for both space and the ground.

Each lecture was first given in Boston and then repeated the following week in Washington, DC, under the aegis of the Smithsonian Institution's Resident Associates Program. (The scheduling of speakers for appearances on two consecutive weeks posed some unique logistical problems. Four of the six speaker-authors are 'observational cosmologists,' that is, their theoretical arguments are often based on observational evidence they have gathered themselves. Since the telescopes at major observatories in Arizona, Hawaii, and Chile are seriously oversubscribed, few astronomers are willing to give up precious time during dark-sky periods. Thus, this may be one of the few lecture series scheduled around phases of the Moon!)

The Lowell Lectures at the Boston Museum of Science were coordinated by John Carr, then head of the Charles Hayden Planetarium and one of the founders of this highly successful series in 1975. As always, the members of the planetarium staff provided valuable assistance in everything from handling advance reservations to ushering on lecture nights. The Washington segment of the series was handled by Anna Caraveli, Program Coordinator of the Resident Associates Program. The original idea of sharing these lectures with a

Washington audience came from Ross Simons of the Smithsonian. At the Center for Astrophysics, I am particularly grateful for the help of John Hamwey and Steve Seron of the Publications Department, who provided, respectively, illustration and photographic support. Mary Juliano, department secretary, retyped major portions of the manuscript and provided administrative assistance for several aspects of the lectures and book preparation. Various members of the Planetarium Advisory Committee, and especially Bart Cardon, offered valuable advice and counsel in the early stages of lecture planning. Finally, my special thanks go to the speaker-authors, each of whom gave a magnificent lecture and then supplied me with a near-perfect written version for publication. This is their book in every sense.

Cambridge, Massachusetts   James Cornell
January 1988

DISTANCES TO HIGH REDSHIFT OBJECTS

Directly measuring distances to celestial objects is one of the greatest challenges in astronomy. For nearby objects one can use geometrical techniques, just as does a surveyor. However, for all but the nearest stars, and certainly for all the objects that lie outside of our galaxy, such methods are inadequate. Fortunately, for distant objects such as quasars and galaxies, the universe provides us with a convenient indirect method of finding distances. Because the universe is uniformly expanding, objects that are more distant appear to be receding from us at greater speeds than their nearby counterparts. Thus, by measuring how fast a galaxy appears to be receding from us, we can calculate its distance.

The speed at which an object is receding from us is determined from the shift in its spectrum. As the universe expands, the distance between objects is stretched. Light rays traveling between them are also stretched to longer wavelengths, and hence become redder. The degree to which the light is stretched is expressed as a *redshift*, the fractional amount by which the light has had its wavelengths increased. For example, a galaxy or quasar with a redshift of 1 has had the wavelengths of its light increased by 100 percent, and a quasar with a redshift of 4 has had its wavelengths increased by 400 percent. Redshift is represented by the symbol z.

Converting a redshift into a distance (that can be expressed in units of miles, for example) requires knowledge of the expansion rate of the universe. Unfortunately, the rate at which the universe is currently expanding is uncertain. The expansion rate, called the *Hubble Constant*, is usually expressed in units of kilometers per second (a speed) per megaparsec (a distance). In these units, the value of the Hubble Constant is believed to be between 100 and 50. In the more familiar units of miles per hour per million light years, the value of the Hubble Constant is in the range 70 000–35 000 mph/mly. In other words, a galaxy that is 1 million light years away from us will appear to recede from us at a speed of 35 000–70 000 miles per hour.

For very large distances there is a further complication. If we look at a galaxy 1 million light years away, then, by definition, it has taken

xi

the light 1 million years to reach us. Thus, if we take a picture of this galaxy, we are actually seeing it as it was when the light left it, some 1 million years ago. When we look at galaxies and quasars at larger and larger distances, we are actually looking farther and farther back in time. The difficulty in calculating distances comes about because the rate at which the universe is expanding may have been different in the past than it is today. This changing of the rate of universal expansion is characterized by a number called the *deceleration parameter*, represented by the symbol q_0. The value of this parameter is believed to lie in the range between 0 and 0.5. A universe in which $q_0 = 0$ is said to be 'open' and expands at a constant rate; a universe in which $q_0 > 0.5$ is said to be 'closed,' and its rate of expansion is decreasing and some day will stop altogether.

Both the table and the figure that follow give the redshifts for a number of astronomical objects, including recently discovered quasars with redshifts of 4 and greater. For each redshift, the distance is listed in billions of light years – as it would apply to different cosmologies. For example, the table lists distances for both 'open' and 'closed' universes, and for values of Hubble's Constant at values of 100 and 50. In the graph, however, the distances are plotted only for $H_0 = 50$; to obtain the distances for $H_0 = 100$, simply divide by 2. (Adapted from material prepared by Patrick J. McCarthy, University of California, Berkeley, for the American Astronomical Society.)

Distances to celestial objects for various cosmologies in billions of light years

Redshift	$H_0 = 50$ $q_0 = 0$	$H_0 = 50$ $q_0 = 0.5$	$H_0 = 100$ $q_0 = 0$	$H_0 = 100$ $q_0 = 0.5$	Objects
0.001	0.02	0.02	0.01	0.01	Nearby galaxies
0.50	6.52	5.94	3.26	2.97	Clusters of galaxies
1.00	9.78	8.43	4.89	4.21	Radio galaxies
1.80	12.57	10.25	6.28	5.12	Radio galaxies
3.00	14.67	11.41	7.33	5.71	Quasars
4.01	15.62	11.87	7.82	5.94	Quasar
4.10	15.72	11.91	7.86	5.96	Quasar
4.30	15.86	11.97	7.93	5.98	Quasar
4.40	15.93	11.99	7.96	5.99	Quasar
4.43	15.95	12.01	7.97	6.01	Quasar
5.00	16.29	12.15	8.14	6.07	?
∞	19.55	13.04	9.77	6.52	Big Bang

Redshift vs Distance

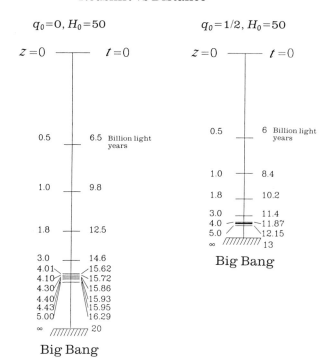

1

Discovering the universe: an introduction

ALAN P. LIGHTMAN
*Harvard–Smithsonian Center for Astrophysics
and
Massachusetts Institute of Technology*

Introduction

An ancient Assyrian cylinder, dating back to 700 BC, shows a worshipper standing between two Babylonian gods (Figure 1.1). The god on the left, holding an ax, is thought to be Marduk, who played a major role in one of the oldest recorded stories of creation, the Enuma Elish. According to the Enuma Elish, the universe began as a watery chaos. Eventually, the heaven and earth gods were born, but imprisoned within the body of Ti'amat, the goddess of chaos. Marduk does battle with Ti'amat, kills her, and with his ax cuts her body in two. One half of her body he lifts up to form the heavens. The other half becomes the Earth.

The Enuma Elish exemplifies the strong anthropomorphic quality of early cosmological myths. It also portrays the human desire for order and stability, a repeating theme in subsequent cosmologies. Although cosmological speculations have gradually shifted from myth to physical theory, they have always embodied world views, and as such have partly reflected the religious, psychological, and philosophical ideas of the time. By the same token, changes in cosmological thinking have rarely been brought about simply by new technological developments or by new observations of the facts of nature.

For the moment, let us jump 3500 years forward from the Enuma Elish to sketch our current view of the universe. The study of the structure and evolution of the universe as a whole is a subfield of astronomy called cosmology. We now believe that the basic unit of structure on the cosmological scale is the galaxy, an isolated congrega-

Figure 1.1 Assyrian cylinder, showing a worshipper between two divinities. The god on the left, holding the ax, is probably Marduk. The goddess, wearing the crown, is probably Ishtar. (Photograph from Bibliothèque Nationale, Paris.)

tion of about 100 billion stars. A typical galaxy is about 100 000 light years in diameter. (A light year is about 6 trillion miles, which is the distance a light beam travels in a year.) Our own galaxy, the Milky Way, is thought to be very similar in appearance to the Andromeda Galaxy (Figure 1.2). Andromeda is relatively nearby at a distance of about 2 million light years away. On average, galaxies are separated by 10–100 galaxy diameters.

In the 1920s, we learned that the universe is in motion. It is expanding, with all the galaxies moving away from each other. When this expansion is extrapolated backwards in time, the galaxies move closer and closer to each other, until a definite point in the past when all the matter in the now observable universe was crammed together into a region of space smaller than an atom. This point of time, called the Big Bang, happened about 10–20 billion years ago, according to our best estimates. The Big Bang marked the birth of the universe as we know it. Thus, the universe is not only constantly changing, but it also seems to have had a beginning. Almost none of these facts were known as recently as 100 years ago. (Of course, medieval scholars would have said they knew the universe had a beginning, based on religious and philosophical considerations.)

Figure 1.2 Andromeda Galaxy, about 2 million light years away. (Harvard College Observatory photograph.)

Cosmology is a vast subject. I'll confine my discussion here to the history of just two scientific themes: distance, and the overall motion of the universe. In parallel, I will consider two philosophical and psychological themes: the human desire for order and permanence, and our changing view of our place in the universe. First, we thought the heavens were divine and centered on us. Then, we gave up being the center but held to the belief that the heavens didn't change. Finally, we have come to realize that everything in nature, including the heavens, is made of the same perishable stuff. Our discovery of the universe has been, in part, a discovery of ourselves.

The Aristotelian universe

Aristotle constructed the world out of five elements: earth, water, air, fire, and aether. Everything had its natural place. The natural place of Earth was at the center of the universe, and all earth-like particles drifted to that location. Aether was a divine and indestructible substance; its natural place was in the heavens, where it made up the stars and the other heavenly bodies. Water, air, and fire had intermediate locations.

The Aristotelian world view is partly illustrated in the medieval painting 'Expulsion of Adam and Eve from Paradise', by Giovanni di Paolo (Figure 1.3). The motionless Earth sits at the center of the

Figure 1.3 'Expulsion of Adam and Eve from Paradise', by Giovanni di Paolo, depicting the medieval world view. (From the Robert Lehman Collection, The Metropolitan Museum of Art, 1975.1.31.)

universe. The Sun, planets, and stars are attached to rigid, crystalline spheres, which revolve in perfect circles about the Earth. The outermost sphere, the *primum mobile*, is spun by the finger of God. The inner spheres rotate, in sympathy, for the love of God. Aside from the revolutions of the heavenly spheres, Aristotle's universe is static. It is eternal, without beginning and without end.

Aristotle's cosmology was later modified by letting the Sun and planets revolve in small circles whose centers revolved in large circles about the Earth. In this way, the distance from a planet to the Earth could change in time, as required by the observed changes in brightness of the planets, while still maintaining the two notions that the Earth sat at the center of the universe and that the orbits of heavenly bodies were composed of perfect circles. The most influential of these modified Aristotelian systems was set forth by Ptolemy (*ca.* 100–170 AD), in 13 volumes called the *Almagest*. Ptolemy was also the author of the *Geography*, the leading map of the known world for many centuries.

The early Greeks did more than speculate and philosophize about cosmology. They also made concrete measurements. It was the Greek Eratosthenes who first accurately determined the radius of the Earth, in 196 BC. His method is illustrated in Figure 1.4. Because the Earth is curved, the direction to the Sun at a given instant will vary, depending on where you are on the Earth, and this variation may be used to measure the radius of the Earth. Eratosthenes knew that at the city of Syene (now Aswan), the Sun was directly overhead at noon on the first day of summer, because at that moment the Sun's rays lit up the bottom of a deep vertical well. At the same time, the Sun's rays made an angle of 7 degrees, or 7/360 of a circle, with respect to the vertical at Alexandria, 500 miles away. Assuming that the Sun's rays come in almost parallel, then simple geometry tells us that the distance between Alexandria and Syene is 7/360 of the total circumference of the Earth. This yields a value of about 25 000 miles for the circumference of the Earth, close to the best modern value. Although accurate and stable clocks were well beyond Greek technology, it was not necessary to have synchronized clocks at Syene and Alexandria in order to measure the direction to the Sun at the same instant. Since Alexandria is mostly north of Syene, a fact that could be determined by the Greeks, Eratosthenes needed only to measure the *minimum*

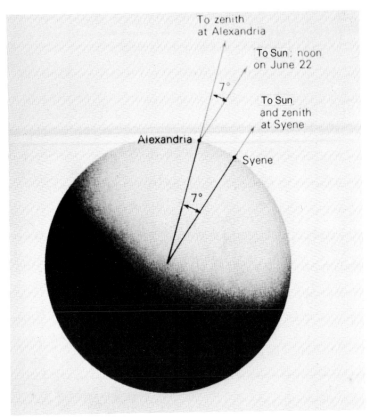

Figure 1.4 Eratosthenes' method of measuring the size of the Earth.

angle between the Sun's direction and the vertical at Alexandria during the first day of summer. Such a minimum would occur approximately when the Sun was overhead at Syene.

The Greeks also measured larger astronomical distances. However, aside from the distance to the Moon, these distances were all highly underestimated. For example, the Greeks figured that the Sun's distance was 200 Earth diameters, when, in fact, it is about 12 000. Good estimates of the dimensions of the solar system required instruments with far more precision than those available to the Greeks. It would be almost 1900 years before such instruments could be made and used.

The Copernican universe

Nicolas Copernicus (1473–1543), a Polish astronomer, was largely responsible for overthrowing the Earth-centered, Aristotelian world view. Copernicus placed the Sun at the center of the solar system and relegated the Earth to one of the planets, orbiting about the Sun. This new system didn't fit the observations any better than the Aristotelian–Ptolemaic system, but it was a lot simpler.

Figure 1.5 Nicholas Copernicus. Thought to be a self-portrait. (Courtesy of the Municipal Museum of Torun, Poland.)

At the beginning of his great masterwork, *On the Revolutions of the Heavenly Spheres* (*De Revolutionibus Orbium Coelestium*) (1543), Copernicus gives credit to some of the early Greeks for the idea of a moving Earth. However, the idea had more impact when Copernicus proposed it, even before supporting observations by Galileo's telescope, 70 years later. Copernicus lived in a time of religious and cultural change. The absolute authority of the Roman Catholic church was being questioned, by Martin Luther in Wittenberg and by John Calvin in Geneva; Henry VIII was arguing with the Pope, although Henry had things on his mind other than cosmology. Perhaps in this environment, people were more susceptible to new cosmological views. Another factor in the receptiveness to a new map of the heavens might have been the radical revisions in maps of the Earth. Ptolemy's ancient *Geography*, which had been based in part on the reports of sailors and merchants, was grossly inaccurate beyond the walls of the Roman Empire. The new geographical knowledge of the fifteenth and sixteenth centuries, gleaned in part from the voyages of Columbus and others, undermined confidence in Ptolemy as a reliable map-maker and thus cast doubt on his astronomical system as well.

The removal of man from the center of the universe was difficult for scientists as well as for theologians. It was widely believed that the universe was made for man's benefit and that human beings were special. For example, the German astronomer Johannes Kepler (1571–1630) accepted the Copernican system, but still emphasized that our Sun was the center of the universe. In Kepler's *Conversations with Galileo's Sideral Messenger* (1610), Kepler points out that, according to his calculations, our Sun is the most luminous (and therefore the noblest) star in the Milky Way. For Kepler, this is perfectly reasonable since:

> . . . if there are globes in the heaven similar to our Earth, do we vie with them over who occupies the better portion of the universe? For if their globes are nobler, we are not the noblest of rational creatures. Then how can all things be for man's sake? How can we be the master of God's handiwork?

(trans. New York: Edward Rosen, 1965, page 43)

Although Copernicus changed many things, he left some things unchanged. Figure 1.6, from *On the Revolutions*, shows Sol at the center of the universe with the orbits of the various planets about Sol.

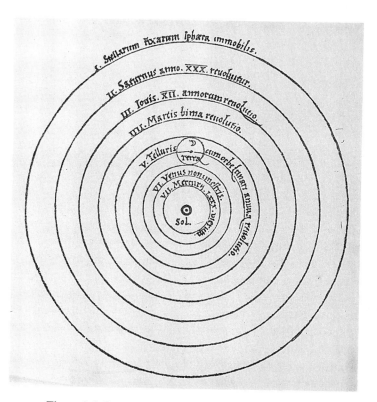

Figure 1.6 Copernicus' system of the universe. From *On the Revolutions*.

Terra is in third position, after Mercury and Venus. Note that the orbits of the planets are still perfect circles, as in the Aristotelian system. Note also that the outermost circle is labeled as the 'Stellarum Fixarum Sphaera Immobilus,' the 'immobile sphere of the fixed stars.' Like Aristotle, Copernicus continued to believe that the stars were fixed and unchanging, even divine. He explains his view in this way (*On the Revolutions*, I, cap VIII):

> The condition of being at rest is considered as nobler and more divine than that of change and inconsistency; the latter, therefore, is more suited to the Earth than to the universe.

Here Copernicus contrasts the constancy of the heavens to the impermanence of the Earth, a contrast that has been used repeatedly in literature, sometimes with religious associations. In Percy Shelley's long poem 'Adonais' (1821), for example, are the lines:

> The One remains, the many change and pass,
> Heaven's light forever shines, earth's shadows fly.

The Newtonian universe

The next major revolution in cosmological thought might be associated with Isaac Newton. The Newtonian world view placed all phenomena of nature on an equal footing – including the stars. The groundwork for the Newtonian universe was laid by an Italian philosopher Giordano Bruno (1548–1600) and an English astronomer Thomas Digges (1546–1595). Bruno and Digges independently proposed that the universe was infinite, with the stars scattered outward through infinite space rather than attached to an outermost heavenly sphere. Bruno further proposed that the stars were suns, perhaps with their own planets and intelligent life. (For this, and other indiscretions, Bruno was burned at the stake.) These new ideas took attention away from the planets and focused it on the stars. More importantly, the stars had been pried loose from their crystalline spheres. Stars were now physical objects, like planets. They would be subject to the same physical laws.

The notion of universality of physical laws, applying in the heavens as well as on Earth, was given its most brilliant exposition by the French philosopher and mathematician Rene Descartes (1596–1650) (Figure 1.7). Descartes' great work, *Principles of Philosophy* (1644), was the most exhaustive study of nature and philosophy undertaken since Aristotle's. In *Principles* Descartes proposed that the universe was not made for man, a new idea. Descartes went on to compare the universe to a giant clock, obeying mechanical laws. Newton was much influenced by Descartes, as we will see. But first, let us review some of the technical developments, which we left in 200 BC with Eratosthenes' measurement of the size of the Earth.

In 1610, Galileo Galilei (1564–1642) put into use the first high-tech instrument in astronomy, the telescope (Figure 1.8). With his telescope, Galileo resolved the 'cloud' of the Milky Way into individual stars, discovered craters on the Moon, and saw other moons around Jupiter. This latter discovery rebutted a leading argument in favor of

Figure 1.7 Rene Descartes. (Courtesy of Owen Gingerich.)

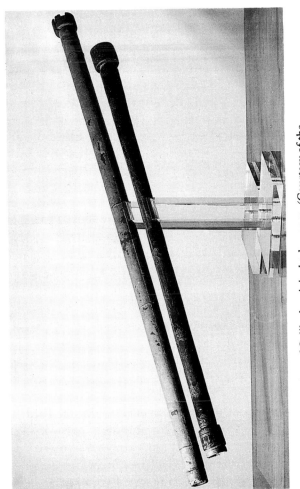

Figure 1.8 Two of Galileo's original telescopes. (Courtesy of the Museum of History of Science, Florence.)

an Earth-centered cosmos. If other planets could have bodies revolving about them, then the Earth could not be unique in this respect. Galileo also made the discovery that some of the misty, white patches long seen in the sky were actually collections of stars, just too far away to be resolved individually with the naked eye. Later, some of these patches, called nebulae, were found to be entire galaxies.

The telescope allowed the first accurate measurement of the dimensions of the solar system. This measurement was accomplished in 1672 by Jean-Dominique Cassini (1625–1712), an Italian astronomer who came to Paris to work (and was given a rough time until he became fluent in French). Actually, Cassini measured the distance to Mars, but from this the other solar system dimensions could be determined, since the ratios of solar and planetary distances had long been known from the observed orbital periods and motions of the planets. Cassini used a now familiar technique known as the parallax method. Two observers, in two widely separated locations, simultaneously measure the position of the planet. Because of their different locations, the two observers have different viewing angles and so will measure slightly different positions of the planet relative to the background of stars. The effect is similar to that when you look at a tree in your front yard, then walk a few feet to the right and look at the tree again. The position of the tree shifts a little relative to the neighbor's house across the street. From the amount of the angular shift and the distance between the two observation points, you can measure the distance to the tree. In Cassini's measurement, one observation point was in Paris and the second was in Cayenne, in South America. (Cassini stayed in Paris; his assistant was sent to Cayenne.)

An interesting obstacle to using this method is that the Earth is moving through space. Consequently, if the two observations of Mars aren't made at the same instant, the distance between the two observers cannot be easily determined. To make the two measurements simultaneously, the two observers must have accurately synchronized clocks. Unfortunately, such clocks didn't exist. The best clocks in the seventeenth century were pendulums. Although two pendulums might be initially synchronized, they certainly wouldn't be after one was transported 5000 miles away, including a bouncy voyage across the ocean. In 1666, Cassini discovered an ingenious remedy to the problem of synchronizing clocks. Using the telescope, he saw that the moons of Jupiter could be used as universal clocks in the sky. Each moon, as it orbited behind the planet, reappeared at a precise instant of time. Thus two observers at any two places on Earth within view of Jupiter could set their clocks by the reappearance of one of its moons.

After Cassini's measurement in 1672, the size of the solar system was at last known. It would be more than 150 years before astronomers directly measured the distance to the nearest star (using a different version of the parallax method). However, Isaac Newton (1642–1727)

Figure 1.9 Portrait of Isaac Newton. (Courtesy of Owen Gingerich.)

(Figure 1.9) *estimated* the distance to the nearest stars, by *assuming* that their intrinsic luminosities were the same as the Sun's. Then, by comparing their apparent brightnesses to the Sun's, he could determine the stars' distances, which came out to about 5 light years. Although this is a pretty good estimate, Newton's method cannot be used in general, since stars have a fairly large range of intrinsic luminosities.

Isaac Newton's masterwork was the *Principia*. The *Principia*, with its theory of gravity and its laws of motion, gave a rigorous mathematical foundation to Descartes' notion of the universe as a giant, mechanical clock. Figure 1.10 shows the title page of the first edition of the *Principia*, published in 1687. Note the obvious influence of

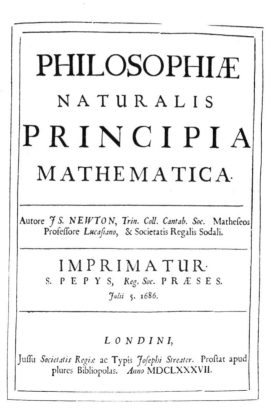

Figure 1.10 Title page of Newton's *Principia*. (Courtesy of Owen Gingerich.)

Descartes. Two words from the title, *Philosophiae Principia*, exactly repeat the title of Descartes' great work.

Newton correctly realized that gravity is the most important force for understanding the large-scale structure and behavior of the universe. (It is possible that local, non-gravitational forces also play a role in cosmology, as suggested by recent theoretical ideas in particle physics and recent observations of the distribution of galaxies. See the chapters by Alan Guth and Margaret Geller.) However, Newton was a highly religious man, as well as a great scientist, and he equated space to the body of God:

> the Supreme God is a Being eternal, infinite, absolutely perfect . . .
> He endures forever and is everywhere present; and by existing
> always and everywhere, he constitutes duration and space
>
> (*Principia*, General Scholium, translated in the *Principia*, Berkeley:
> University of California Press, 1962, vol. 2, page 544).

Elsewhere in his writings, Newton associated constancy and stability with the perfection of God, and change with friction and decay. In his *Opticks*, for example, Newton says that 'Motion is much more apt to be lost than got, and is always upon the Decay', and that irregularities in planetary orbits 'will be apt to increase, till this System wants a Reformation' from God (*Opticks*, New York: Dover, 1952, pages 398 and 402). Perhaps for these reasons, Newton claimed that the universe must be static, without attempting to explore the problem quantitatively. Newton's argument for a static universe did not appear in his published works, but rather in a letter sent to the theologian Richard Bentley in 1692. The argument runs as follows: If the universe were globally expanding or contracting, there would have to be a center to the motion; but matter scattered uniformly through an infinite space does not define a center – therefore the universe must be static. In fact, Newton's own equations predicted that the universe had to be in motion, either contracting or expanding. The error in Newton's argument, as we will see later, is that global motion does not require a center.

The association of evolution with friction and disorder was continued in the next century in the cosmologies of Thomas Wright (1711–86), Johann Lambert (1728–77), William Herschel (1738–1822), and Immanuel Kant (1724–1804). We are most familiar with Kant as a pure philosopher, but he also indulged in some serious cosmological speculation. In Kant's cosmology, *Universal Natural History and Theory of the Heavens* (1755), the gravitational attraction of stars for each other was exactly balanced by orbital motions of stars. Thus, although individual stars had motion, these motions were arranged in a highly systematic way, to keep the entire system in an overall state of balance and constancy. Without such a balance, Kant describes the resulting evolution as leading to destruction and chaos. Furthermore, he points out that a universe not so balanced would lack 'the character of that stability which is the mark of the choice of God' (translated by W. Hastie in *Kant's Cosmology* and reprinted in *Theories of the Universe* (Editor M. K. Munitz), New York: The Free Press, 1957, page 241). Note that God is again invoked as the author of order and stability in the universe. Indeed, scientific papers on cosmology continued to mention God in this context until sometime in the middle to late 1800s. In my opinion, the studied lack of references to God after this time resulted from a change in social convention among scientists rather than any change in underlying thought.

One footnote might be added to our discussion of the popular belief in an intrinsic order and stability in nature. The structure of the solar system was sometimes taken as a small-scale model of the structure of the universe as a whole. In a paper presented in April 1788 to the French Academy of Sciences, the French mathematician and astronomer Pierre Laplace (1749–1827) gave a mathematical proof of the stability of the solar system. He then claimed:

> this stability exhibits in the heavens the same intention to maintain order in the universe that nature has so admirably observed on Earth for the sake of preserving individuals and perpetuating species.
>
> (*Ouevres completes de Laplace*, XI, 248–9; translated in *Dictionary of Scientific Biography*, New York: Scribner's, vol. 15, page 333).

Laplace's admiration of order in the heavens and on Earth came just one year before the French Revolution.

The Einsteinian universe

In 1915, Albert Einstein (1879–1955) (Figure 1.11) put forth a new theory of gravity, the first new theory of gravity since Newton's. Two years later, Einstein applied his theory to cosmology. In his 1917 paper, Einstein assumes from the beginning that the universe is in a

Figure 1.11 Albert Einstein. (Photograph by Johan Hagemeyer, Bancroft Library. Courtesy of the American Institute of Physics.)

state of static equilibrium and then searches for the conditions required to maintain such an equilibrium. He first makes an analogy between a Newtonian system of stars and a gas of molecules: 'If we apply Boltzmann's law of distribution for gas molecules to the stars, by comparing the stellar system with a gas in thermal equilibrium, we find that the Newtonian system cannot exist at all' (translated in *The Principle of Relativity*, by H. A. Lorentz, A. Einstein, H. Minkowski, and H. Weyl, New York: Dover, 1952, page 178). What Einstein means is that an isolated system of stars cannot remain in equilibrium, but will gradually lose stars into distant space, just as water slowly evaporates into the surrounding air. Since Einstein implicitly assumes that our stellar system has lasted an *infinite* period of time, it *must* be in a nonevolving equilibrium state; any process that might destroy it, no matter how slowly, would have done so already. Thus, the Newtonian system 'cannot exist at all.' Einstein never considers the possibility that the universe might be of finite duration and out of equilibrium. He then discusses a modification of Newton's theory that allows a stellar system to maintain a static equilibrium and extends this modification to his own new theory. Eventually, Einstein obtains static (time-independent) solutions to his equations.

Why did Einstein assume that his cosmological model had to be static? This interesting question requires a small digression. From the papers and personal letters of astronomer Wilhelm de Sitter in 1917, Einstein would have known that astronomical observations *did not require* a static universe. Furthermore, the strong religious motivations for a static universe found in Newton's writings are absent in Einstein's. We can only speculate as to Einstein's thinking. One possibility is simply that a static cosmology is less complex than an evolving cosmology. Another possibility is that Einstein, like other scientists before him, saw one-way change as disintegration and hence philosophically and psychologically objectionable. In his 1917 paper, Einstein says that without proper conditions imposed on the external gravitational field, radiation and stars would escape the stellar system, becoming 'ineffective and lost in the infinite' (*The Principle of Relativity*, page 178). But with the proper external gravitational field, introduced by a modification of the theory, the stellar system 'would not therefore run the risk of wasting away' (*The Principle of Relativity*, page 179). Einstein's thermodynamic language, and especially the phrase 'wasting away', resonates with the second law of thermodynamics, discovered in the previous century. The second law of thermodynamics states that any closed system will inevitably become more disordered in time and eventually run down. When applied to the universe as a whole, this law has unappealing implications, and many noted scientists of the last 150 years have resisted these implications. For example, William Rankine (1820–72) proposed that giant reflecting walls in distant space somehow captured and refocused into usable form the energy lost by decaying systems. And William

Thomson (a.k.a. Lord Kelvin; 1824–1907), one of the discoverers of the second law of thermodynamics, stated in 1862 that it was

> impossible to conceive a limit to the extent of matter in the universe; and therefore science points rather to an endless progress . . . than to an single finite mechanism, running down like a clock, and stopping forever
>
> (*Popular Lectures and Addresses*, London: Macmillan, 1891, vol. I, pages 349–350).

As late as 1928, the Nobel Prize winning physicist Robert Millikan argued that his theory of the creation of atoms in space refuted the concept of a universe falling apart in the grips of the second law of thermodynamics:

> With the aid of this assumption one would be able to regard the universe as in a steady state now, and also to banish forever the nihilistic doctrine of its ultimate heat death
>
> (*Science and the New Civilization*, New York: Scribner's, 1930, pages 108–9).

These views and comments suggest psychological (as opposed to purely scientific) objections to a universe 'wasting away,' and it is conceivable that Einstein had similar feelings. In any event, as we will see, he not only proposed a static universe but also resisted the notion of an evolving universe.

In 1922, the Russian mathematician and physicist Alexander Friedmann found solutions to Einstein's cosmological equations that evolved in time, describing an expanding or contracting universe. At the beginning of *his* paper, Friedmann points out that Einstein's implicit assumption of a static universe is indeed only an assumption and not required by any observations. During the next year, Einstein published two short replies to Friedmann's paper. In his first reply, Einstein claimed that Friedmann had made an error in calculation and that 'the significance of [Friedmann's] work consists exactly in the fact that it proves this time independence' (*Zeitschrift fur Physik*, 1922, **11**, 326). In his second reply, Einstein acknowledged that Friedmann's calculations were, in fact, correct and 'clarifying,' presenting an alternative to his own earlier static cosmological model (*Zeitschrift fur Physik*, 1923, **16**, 228). However, in the hand-written draft of this paper is a crossed-out sentence fragment saying that, to Friedmann's time-dependent solution of the cosmological equations, 'a physical

Figure 1.12 Edwin Hubble. (California Institute of Technology photograph.)

significance can hardly be ascribed' (*The Collected Papers of Albert Einstein*, unpublished document 1-026; translated and quoted with the permission of the Hebrew University of Jerusalem, Israel). Einstein thus rejected Friedmann's time-dependent solutions as being unphysical. Six years later, in 1929, the American astronomer Edwin Hubble (1889–1953) (Figure 1.12) found conclusive evidence that the universe is indeed expanding and changing in time. Enough for Einstein's belief in a static universe.

To understand Hubble's discovery, we have to go back some years.

It had been known since the early 1900s that many of the nebulae, those distant misty patches of stars, were in motion, speeding away from the Earth. The outward motion of the nebulae was determined by a technique known as the Doppler shift. When a source of sound or light is in motion, its frequency changes, becoming higher if the source is moving toward you and lower if it is moving away. The whistle of a train rises in pitch when the train is approaching and drops in pitch when the train is receding. In light, the analogue of pitch is color. Thus, the color of an approaching light bulb shifts toward the blue end of the spectrum, toward higher frequencies; the color of a receding light bulb shifts towards the red end of the spectrum. Unless the source of light is moving at near the speed of light, these shifts are too small to detect with the naked eye, but sensitive instruments can detect them. Systematic evidence for the 'redshift' of colors of nebulae was first provided by Vesto Melvin Slipher (1875–1969), in a program of observations extending from 1912 to the mid-1920s and carried out at the 24-inch telescope at the Lowell Observatory, in Flagstaff, Arizona.

The significance of Slipher's result, however, was not clear, since no one knew what the nebulae were. Were they nearby or far away? How luminous were they? These questions could be answered if the distances to the nebulae were known, but it is extremely difficult to determine distances in astronomy, as previously discussed. A major problem is that stars come in a wide variety of intrinsic luminosities. Thus a star that appears bright could either be nearby, and of modest intrinsic luminosity, or far away, and of large intrinsic luminosity. In the second decade of the twentieth century, a technique was discovered by Harvard astronomer Henrietta Leavitt (1868–1921) for determining the intrinsic luminosity of certain stars, called Cepheid variables. Leavitt discovered that Cepheid variables oscillate in brightness, with the rate of oscillations uniquely determined by the Cepheid's intrinsic luminosity. (She determined this by examining a group of Cepheids all located in the same nebula and thus at the same distance.) By measuring the period of oscillation of a Cepheid's light, you could determine its intrinsic luminosity. By measuring how bright it appeared, you could then calculate its distance. It is as if Cepheid stars were a bunch of light bulbs, with the property that all the 50-watt bulbs flickered at 10 times per second, all the 100-watt bulbs flickered at 17 times per second, and so on. Looking at a bulb in a dark room, you could determine its wattage by seeing how fast it flickers. Then, by noting how dim it appears, you could calculate its distance. The Cepheid variables thus served as 'standardized light bulbs' and became good distance indicators. A Cepheid variable found in any region of space could be used to determine the distance to that region. In this way, Cepheid variables were used by Harlow Shapley (1885–1972) in 1919 to determine the size of the Milky Way.

In 1924, Edwin Hubble happily found a Cepheid variable in the Andromeda Nebula. On measuring its distance, he determined that the Cepheid was far outside our galaxy, thus indicating that the nebulae are entire galaxies, not just small, nearby groupings of stars within our galaxy. Galaxies, not stars, are the basic units of matter in the universe. Furthermore, these other galaxies are in motion, heading away from us. In 1929, using the 100-inch diameter telescope at the Mount Wilson Observatory in California, Hubble found evidence that the outward speed of a galaxy is directly proportional to its distance away from us; galaxies twice as far away are receding from us at twice the speed. This is exactly what would be expected if the entire universe is expanding, as discussed in a theoretical paper by the Belgian astronomer Georges LeMaitre in 1927. To help visualize the manner in which the universe is expanding, imagine that galaxies are dots, painted on the surface of a balloon, as in Figure 1.13. (This analogy was first used by the British astronomer Sir Arthur Eddington in 1931.) If the balloon is expanding, then from the point of view of any one of the dots – for example, our dot, the Milky Way – it appears that all the other dots are moving away; furthermore, the view is identical from *any* dot. There is no special dot. There is no center of the expansion. This was Newton's mistake. He didn't visualize a situation in which all points could be moving away from all other points, but with no center of expansion. (Of course, the *balloon* has a center, but we are only talking about the *surface* of the balloon. The interior of the balloon and the space around the balloon are not part of our two-dimensional analogy). We can also see in this analogy why the speed of recession of a dot is proportional to its distance from our dot. Suppose dots A and B are currently 1 and 2 inches north of our dot, respectively, and dot A is moving north at 1 inch per second. Now analyze the situation from the point of view of dot B, using the

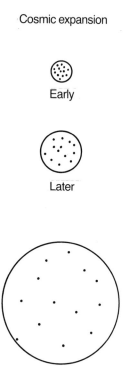

Cosmic expansion

Early

Later

Later

Figure 1.13 Two-dimensional analogy for cosmic expansion. Galaxies are represented by dots on the surface of an expanding balloon.

equivalency of dots. Since dot A is 1 inch south of dot B, it must be moving south at 1 inch per second as seen by dot B. For this to be true, however, dot B must be moving north at 2 inches per second relative to our dot. In conclusion, dot B is twice as far from us as dot A, and it is moving away from us at twice the speed.

By 1931, many astronomers and physicists understood the observations of Hubble to mean that the universe is expanding. This unwelcome fact of nature required a new and unsettling view of the world. A universe expanding was a universe changing. An expanding universe most likely had some kind of beginning, and most likely some kind of end. In the February 1933 issue of *The Atlantic*, journalist George Gray wryly observed that:

> . . . just as the shifting of book-keeping accounts into the red measures disintegrating, scattering, dissipating financial resources, so the shifting of starlight into the red indicates disintegrating, scattering, dissipating physical resources. It says that the universe is running down . . . To entertain this preposterous idea of all these massive star systems [galaxies] racing outward [is] to accept a radically new picture of the cosmos – a universe in expansion, a vast bubble blowing, distending, scattering, thinning out into gossamer, losing itself. The snug, tight, stable world of Einstein had room for no such flights.

The expansion of the universe is probably the most important discovery in astronomy ever made. Many of the critical questions in cosmology today – such as what is the ultimate fate of the universe, why is the universe so homogeneous on the large scale, what kind of grand unified force existed in the first fraction of a second after the Big Bang – would not exist if the universe were static.

In the coming chapters, leading cosmologists will describe what we know about the structure of the universe billions of light years from Earth, and what we know about the first split second after the birth of the universe. Cosmologists today have an outrageous self-confidence. Although fastened to this small planet, we have projected our minds across vast distances of space and time. Ironically, as we have adjusted to an increasingly more humble place in the universe, we have felt an increasing confidence in our ability to understand it. Of course, Aristotle felt that way too. He was just wrong.

Further reading

Munitz, M. K., Editor. *Theories of the Universe*, New York: The Free Press, 1957.

North, J. D. *Measure of the Universe*, Oxford: Oxford University Press, 1965.

2

Measuring the universe: redshifts and standard candles

ROBERT P. KIRSHNER
Harvard–Smithsonian Center for Astrophysics

Distances

In a planetarium, the stars are soft spots of light, some bright and some dim, but all at the same distance, projected onto a dark dome. In the sky, the stars are soft spots of light, some bright and some dim. How do we know they are not fixed to a dark bowl of the night, all at the same distance? The stars are so far away that it takes precise measurement and a little geometrical cunning to tease their distances from their motions. However, once we learn to place the stars at their true distances, we can break out from a flat image of space into a richer three-dimensional view. Starting with the stars and their distances, then sighting the same types of stars in other galaxies, and then learning the properties of these galaxies, we can slowly stock a toolbox that allows us to construct stable foundations for a bridge to the edge of the observable universe.

How do we start? The clues we use to judge distance in everyday life are not all helpful. For example, we know when a tree is in front of a house, it may block part of our view of the porch. Or, we might see a shadow of the tree falling across the wall. But these clues derived from position don't help us judge the distances of points of light. Similarly, we can judge the distance of a person pretty well, because we know the real size of a person. When they're near, they look large; when they're distant, people look smaller. This technique has its parallel in astronomy: once we know what to expect, we can judge the distances to objects by their appearance. But getting on the first rung of the distance ladder can't start this way: we don't know what to expect from stars until we have gauged a few.

26

Another clue we use comes from our stereo vision. The view from your left eye is a little different than the view from your right eye. You can demonstrate this yourself by holding a finger upright in front of your face. If you look with your right eye only, then switch to your left, you will notice that your finger seems to jump position relative to the background. This difference in viewpoints is a useful clue to the distance of objects in the real world, since the amount of shift decreases as the object moves away. Parallax is the name for this shift. Parallax measurements allow us to begin exploring stellar distances.

Strangely enough, it was the absence of parallax that was important in ancient astronomy, just as the absence of a barking dog informed Sherlock Holmes who was molesting the racehorse. The lack of parallax puzzled Greek thinkers as they wondered whether the Earth moved around the Sun, or the Sun around the Earth. If the Earth took an annual tour round the Sun, they reasoned, the stars should look different as our point of view changes. Our view of the stars in a constellation in April should differ from our view of the same set of stars in October due to the parallax shift.

Since there is no obvious distortion of the stars through the year, the Greeks inferred that either the stars were at absurdly large distances (so the shifts were undetectable) or the Earth did none of this unseemly orbiting. The now obvious answer to this puzzle is not that the Earth stands still, but that the distances to the stars are very large compared to the size of the Earth's orbit. However, without a means of measuring the great distances to the stars, many powerful Greek thinkers were convinced that the Earth was at the center of things. In fact, the absence of a measurable parallax shift for stars remained a logical way to refute the Copernican idea even 200 years after it was widely accepted.

Although Cassini had earlier used parallax to measure distances to the planets, measurement of a star's parallax was not possible until telescopes were refined. In 1838, Freidrich Wilhelm Bessel became the first to measure the shift of a star, 61 Cygni.

The parallax angle for the nearest stars is about 1 arcsecond. An arcsecond is a fine old Babylonian unit: a circle has 360 degrees, each degree can be divided into 60 minutes of arc, and each minute into 60 seconds of arc. So an arcsecond is 1/3600 of a degree. If you hold up a finger at arm's length it covers about 2 degrees. (This is roughly true for small people with small fingers and short arms and for large people with large fingers and long arms.) A finger at a distance of 30 meters (100 feet) covers an arcminute, and that's about the limit of human vision. (An eye doctor's chart is labelled 20/20 on the line where the width of the bars in the letter E are separated by about an arcminute.) An arcsecond is the width of a human finger at a distance of about 1500 meters, so it's no wonder that the Greeks didn't notice the parallax of stars. While the stars do demonstrate the parallax effect caused by the Earth's motion around the Sun, it is subtle and demands careful work with telescopes to detect it.

Measuring the parallax of stars is the first step to revealing the size of the universe. We know how big the Earth's orbit is: a substantial 160 million kilometers (93 million miles) in radius. Light traveling from the Sun to the Earth takes about 9 minutes to make the journey, traveling, as it always does, at the speed of light: 300 000 kilometers per second (186 000 miles per second). The nearest star, Proxima Centauri, is so far away that the Earth's orbit covers less than an arcsecond. That corresponds to a distance of 40 trillion kilometers (25 trillion miles). It takes light 4.3 years to travel from Proxima Centauri to us. (Numbers this large, like the national debt or the number of hamburgers sold by McDonald's, lose their impact so it it convenient to think of interstellar distances in units of light years, the distance light travels in 1 year. This slightly odd sounding unit of length has the advantage that the typical distances to the nearest stars are a handful of light years, and at the same time, it reminds us how long it has taken the light from those stars to reach us.)

Measuring the distances to stars helps us place the Sun in perspective. The Sun appears very bright, with enough light scattered by the atmosphere so that we can't see any other stars in the daytime. But a flea thinks a Chihuahua is big, simply because it is so close. Is the Sun an intrinsically bright star or just very near? Parallax measurements show us that many other stars are as bright as our Sun. Indeed, the Sun is a rather common, garden-variety star, not the unique and brilliant object it appears. Having this knowledge is a little like being 12 years old, and slowly coming to realize that your parents are not necessarily the wisest or most wonderful people in the universe, even if they are the closest.

Parallax measurements not only demonstrated to even the most skeptical that the Earth orbits the Sun, but also showed that the Sun was just an ordinary star. The progressive understanding of our place in the solar system, the stars around us, and the galaxies is one in which we find ourselves successively demoted from the center of things, with a special view, to the role of typical participants in the universe, with a view of cosmic events that is not unique. This more democratic view of the universe, in which our vantage point is assumed to be common, rather than special, may be a little humiliating, but it provides a basis for interpreting further measurements of cosmic distances.

Because parallax measurements depend on a delicate measurement of small apparent shifts in position, the range over which they are effective is limited by our measuring precision to about 300 light years. Several hundred stars within that range have well-determined parallax, but they are a small sample of the 10 billion stars in the Milky Way. Moreover, a distance of a few hundred light years is only a very small part of the 30 000 light years to the center of our galaxy. The basic problem is the Earth's atmosphere. If we could get rid of the atmosphere, the images of stars would not shake and dance as they presently do, and we could push our direct surveying of stellar distances out to larger distances. An approach to this problem more practical than removing the atmosphere is to take our measuring tools above the atmosphere. Instruments on the Hubble Space Telescope (HST) and on more specialized satellites are intended to extend this most direct method for measuring stellar distances much farther into the galaxy. (In the last chapter, James Gunn will elaborate further on what we might expect from the HST.)

What is the next step on the staircase to cosmic distances? Learning to recognize types of stars. This approach is similar to other common distance estimates, based on identifying known objects. For example, we can tell a dachshund from a giraffe by its appearance, which doesn't change with distance. Thus we can judge the distance to different animals from a combination of knowing how big they are, and observing how big they seem. Similarly, we learn to sort the stars into different 'species' by observing properties that don't change with distance. These properties include subtle variations in the color and brightness of light emitted by stars. If we can identify a class of stars which are all the same, and we can measure the distance to a few of them by parallax, then we can see whether they all have the same intrinsic brightness. If a class of stars proves to have always the same brightness, then we can use them as 'standard candles' to illuminate the obscure problems of cosmic distances.

Although candles have gone out of fashion for home lighting (and igniting), we sometimes talk about the output of a lighthouse in terms of 'candlepower'. For example, the powerful electric light in a lighthouse is said to have a luminosity equivalent to a million candles of a specified and uniform type – the 'standard candle'. A class of stars can serve as standard candles if all of them have the same intrinsic light output. But the light from a lighthouse is dazzling only if you are close to it, the brightness drops off with increasing distance. Similarly, the apparent brightness of a star decreases as the distance increases. Thus, by measuring the apparent brightness of a star of known intrinsic brightness, we can gauge its distance.

One fortunate shortcut in this painstaking process of establishing a web of distances is that stars, like grapes, form in bunches. Within these star clusters, all the stars are at very nearly the same distance. So, if we find the distance to a handful of stars in a cluster by the reliable method of parallax, or by a well-calibrated standard candle, we also get the distance to hundreds or even hundreds of thousands of other stars, which may include some rare types. This is a good thing, because to get to the largest distances, we need very bright standard candles – and these are most likely to be uncommon stars, with none of them close enough for direct parallax determination.

One especially important type of standard candle is the Cepheid variable. While it may seem oxymoronic to have variable standards, the special properties of these stars make them vital tools for measuring the distances to galaxies. The Cepheids brighten rapidly and then fade slowly, with regular cycles of changing brightness that range from a few days to a few months. Polaris, the North Star, is a Cepheid variable. It changes brightness by about 10 percent every 4 days. Detailed study of Cepheids shows that the stars are actually vibrating, like a weight bouncing up and down on a spring, changing their size and temperature in a regular, repeatable way. Cepheids are distinctive because they are easily identified by their pattern of changing brightness, a property that doesn't change with distance, and one that marks a Cepheid as clearly as the long neck marks a giraffe. Equally important, they are very luminous stars, roughly 10 000 times the brightness of the Sun, and can be detected at very great distances.

One distant place where Cepheids were discovered vibrating away was in the Large Magellanic Cloud (LMC), a nearby companion galaxy to our own Milky Way. South of the Equator, two hazy patches of light are easily seen in the night sky. These are the Large and Small Magellanic Clouds, brought to European attention as a result of Ferdinand Magellan's 1521 voyage around South America. Since stars in the Magellanic Clouds are all effectively at the same distance (which we now know is about 170 000 light years), the stars that appear bright really are luminous, and the stars that appear dim really are dim. The LMC is close enough that many individual stars can be identified and carefully studied; as we shall see, the star that exploded to become the Supernova 1987A had already been carefully examined.

The Harvard College Observatory carried out pioneering studies of the LMC in the opening years of this century (Figure 2.1). The

Figure 2.1 The Large Magellanic Cloud. The nearest galaxy, the LMC is a satellite of our Milky Way. At a distance of about 170 000 light years, it provides an important rung on the ladder of extragalactic distances. This photograph was taken at the Harvard Southern Station in Arequipa, Peru, around 1895. (Harvard College Observatory photograph.)

observatory's Henrietta Swan Leavitt isolated Cepheid variables in the Magellanic Clouds and showed that there was an apparent relation between a Cepheid's period of pulsation and its brightness. This connection between a quantity (the period) that did not depend on distance and intrinsic light output helped weave a net of measurements that could be cast beyond the Magellanic Clouds, deep into the ocean of galaxies, where the big fish, M31 and M33, were lurking.

The distances to these galaxies were measured by Edwin Hubble, using the powerful new 100-inch telescope at the Mount Wilson Observatory. By repeatedly photographing these galaxies in 1923 and 1924, Hubble discovered that they, too, contained variable stars. The pattern of variation could be matched with Cepheids, and their periods measured. Since the relation between period and brightness was known from the Magellanic Cloud sample, the real brightness of these luminous stars could be determined. The amazing fact was that the Cepheids in M31 and in M33 (Figure 2.2) appeared faint: about 100 000 times fainter than you can see with your naked eye, even though they are truly brilliant stars. Just as the Sun is a moderately dim star that appears bright because it is close, the Cepheids in M31 were luminous stars that appeared at the fringe of detectability because they were so distant. The arithmetic shows that the distance to M31 is about 2 million light years.

Once we know the distances to galaxies, we can better appreciate what they are: giant cities of stars with 100 billion individuals bound together by gravity. Hubble's work showed that M31 was not some strange type of cluster in the Milky Way, but actually another galaxy, as big as our own, and that the universe is populated by galaxies, whose nature we are still elaborating and whose origin we still seek.

Cepheids are the most reliable standard candles in our tool kit, but we can only use them out to those distances where individual stars in galaxies can be identified and studied. At present, about 15 galaxies have distances determined from Cepheids. In 1923, the 100-inch telescope was the cutting edge of technology. Today we have new, more powerful and precise detectors on our telescopes that allow measurements of faint objects. These instruments are being used to push Cepheid distance determinations out beyond our local neighborhood of galaxies. The problem of atmospheric blurring crops up again as we strain to measure the pulsations of individual stars in distant

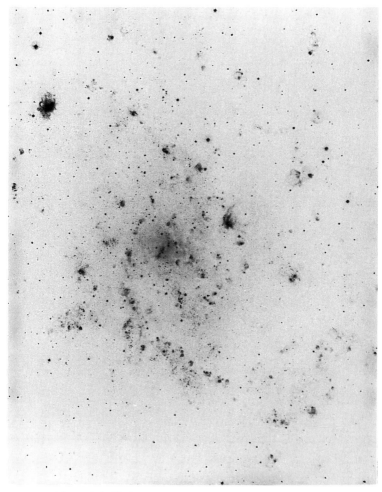

Figure 2.2 M33, a neighboring galaxy. By studying the brightest objects in nearby galaxies, whose distances we can measure by reliable means, we develop tools for use at greater distances. (Robert Kirshner photograph.)

galaxies. The HST will image stars ten times more sharply than a ground-based instrument, so one of its assignments will surely be to search for Cepheids in more distant galaxies. The jackpot in this work will be measuring Cepheids in the Virgo Cluster of galaxies, a gaggle of a thousand galaxies at a distance of about 40 million light years

Figure 2.3 Virgo, Land of the Galaxies. This segment of the sky shows many galaxies of the Virgo Cluster. Clusters such as these are important in measuring the universe because they provide a sample of galaxies which are all at nearly the same distance. (POSS/NGS Sky Survey photograph.)

(Figure 2.3). Just as study of stellar distances is aided by the clumping of stars into clusters, clusters of galaxies help provide samples of galaxies which are all at the same distance.

To increase our range beyond the span where Cepheids serve as yardsticks, we seek brighter objects in the galaxies whose distances we have determined. Or, we attempt to determine properties like the size and brightness of entire galaxies. In the end, we use the brightness of the brightest galaxies as a standard candle to reach out to the edge of the observable universe. These candles are tricky ones, however, because we are looking at light that has traveled 10 billion light years. Therefore, we see the distant galaxies as they were in their youth. Since many of the brightest stars which make these galaxies shine have shorter lives than a few billion years, we will have to learn how galaxies form, develop, and age before we can have confidence that we know how bright they were in the distant past. Comparing dimly seen objects with those nearby that we think are the same may be misleading. Just as we might mistake a model airplane for a real one, and make a large error in distance, our observations of distant galaxies are sketchy and we run some risk of error. This enterprise is difficult, because each step depends on the accuracy of intermediate steps. It is a little like building a tower of blocks: errors at the bottom or misalignments in the middle can result in a very shaky structure at the top. In fact, the current situation is a little embarrassing. There are honest disagreements of a factor of 2 in the distance scale, based on different ways of assessing the long chain of inference that leads to extragalactic distances. That is, some observers claim the most distant galaxies may be 10 billion light years away, but others say 20. It is a big difference, and astronomers are working energetically to make a less elastic yardstick for the largest distances.

One way to put a little backbone into extragalactic distance estimates might be to observe supernovae. The most brilliant supernova in 383 years, SN 1987A, was sighted in February 1987 in the LMC. Supernovae are the brightest stellar phenomenon: a single star blazes as brightly as 10 billion suns for a few weeks, as the center of the star crunches down to the density of a giant atomic nucleus. The outside of the star is ripped apart and heated into a glowing, expanding shell of gas. In the case of SN 1987A, thanks to preexplosion studies of the LMC, we know what star disappeared: it was one called Sanduleak $-69\,202$, a star with 20 times the mass of the Sun. There are two ways that supernovae might prove useful in setting the extragalactic distance scale. One possibility is that supernovae (or, more exactly, one type of supernova) are standard candles: all reaching the same brightness. Another approach is to understand each supernova explosion well enough to estimate its distance, even if each is unique. I have been working on both approaches.

Observing supernovae shows that they fall into two broad classes. The Type I supernovae (SN I) appear to be exploding white dwarf stars, and they look promising as standard candles (Figure 2.4). The problem is that they are rare events. For example, there have been no supernovae identified as Type I in either M31 or the LMC since the invention of telescopes, and the most recent SN I's in our own galaxy were in 1604, 1572, and 1006 AD. We can collect information on Type I supernovae only if we are willing to examine distant galaxies, or sift through the historic evidence. Unfortunately, since we don't

Figure 2.4 A Type I supernova appears. Before and after images of the galaxy NGC 5253, with the bright Type I SN 1972e. In this image, near maximum light, the brightness of a single dying star is comparable to all the other 10 billion stars in the galaxy. Type I supernovae may prove to be good standard candles for surveying the universe. (California Institute of Technology photograph.)

often see them in galaxies at known distances, it is difficult to determine their intrinsic brightness to calibrate them as standard candles. Another approach to finding the true brightness of supernovae is to measure the distance and the apparent brightness of the nearby old events. The distance is the part I've been working on with Frank Winkler at Middlebury College and Roger Chevalier at the University of Virginia, measuring the velocity of the expanding debris at the site of the stellar destruction for the 1572 event and the 1006 explosion. Careful measurement of the motion of those glowing clouds by Sidney van den Bergh and his colleagues can be compared with our velocities to give a distance. We find that the 1006 supernova was about 5000 light years from Earth, and SN 1572 was 7000 light years away.

To get the true brightness, all we need to add is the apparent brightness provided by contemporary observations. For 1604, we have Kepler's records, and for 1572, we have Tycho's to tell us how bright those objects appeared compared to the planets that were visible at the same time (Figure 2.5). For 1006, the Chinese, Korean, Japanese, Arabic, and even European records give clues, but interpreting them is not so easy. For example, the star was said to have 'pointed rays shining so brightly that one could see things clearly'. You would hate to have the modern measurement of the size of the universe hinge on your interpretation of these texts! Still, the idea that Type I supernovae might prove the most powerful standard candles is worth exploring. As we add to the sample of galaxies with good distances, we may find more suitable ways to calibrate the distance scale for SN I.

The 1987 supernova in the LMC was a Type II supernova, the type that is thought to come from a massive star. The identification of the Sanduleak −69 202 progenitor with a 20 solar mass star marked the first time that we've been able to identify the type of star that exploded. But SN 1987A was much fainter than most SN II, because Sanduleak −69 202 was a more compact star than most that become SN II. It illustrates that there is a wider range of brightness for SN II, so they make poor standard candles. But there is hope that by measuring the energy from the supernova, combined with its temperature, and the velocity at which the star is flying apart, we may be able to estimate the distance directly to each individual SN II. This would be just as useful in constructing the dimensions of the universe, and the preliminary work I have done on a handful of supernovae looks promising. The LMC SN 1987A provides a good check for this expanding star method. Since we know the distance to the LMC with reasonable precision, if the supernova measurement yields a different result, the approach may need some further thought. Preliminary results look good for SN 1987A. The great advantage of this method is that it doesn't depend on the parallax of any star, or any other standard candle. It skips the whole elaborate scaffolding of the standard approach to extragalactic distances and provides independent estimates of the distances to the galaxies where we can make good observations of SN II. This method does have one disadvantage. Since it requires rapid response to unpredictable events, it means that carefully crafted telescope schedules need to be juggled in a hurry!

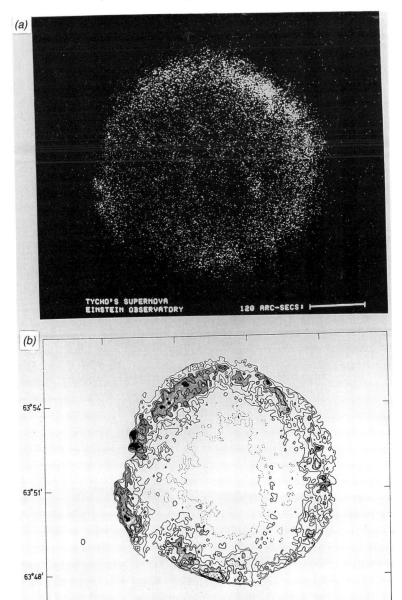

Figure 2.6 CTIO 4-meter telescope. The development of first-rate observing facilities in the southern hemisphere, such as CTIO, has provided powerful new tools for the study of the Milky Way, the Magellanic Clouds, and the galaxies beyond. This instrument was used by a Smithsonian Astrophysical Observatory team to measure the diameter of SN 1987A. (Photograph courtesy of William P. Blair.)

Figure 2.5 Tycho's supernova. At the site of the 1572 explosion of a Type I supernova observed by Tycho Brahe, we now see only a faint ring of optical emission. However, in both X-ray (*a*) and radio (*b*) waves, the expanding 'bubble' of hot gaseous material ejected into space is a striking image. Detailed study of the emission and the motion of the ring allow an estimate of the distance to the Type I, and show one path to setting the extragalactic distance scale. (Smithsonian Astrophysical Observatory illustration.)

An interesting new development in this work comes from the novel measurements being carried out with the optical speckle interferometry technique by Peter Nisenson, Costas Papaliolios and their colleagues at the Harvard–Smithsonian Center for Astrophysics. When the supernova erupted, I called Cos immediately, because I knew he had been working on methods for measuring the sizes of stars that go below the limits set by the atmosphere. Could he make a direct measurement of the diameter of the expanding supernova shell in the LMC? A little arithmetic convinced me that the answer was no, but Cos quite correctly arranged to make the observations anyway at the 4-meter telescope of the Cerro Tololo Interamerican Observatory (CTIO) in Chile (Figure 2.6). They found two very surprising results. One is that the supernova appeared to have an enigmatic and inexplicably bright companion. The other is that the diameter as measured was much larger than expected from simple estimates. While the situation is not clear, it shows that we need to understand the nature of the objects we observe when we try to use them for distance measurements.

Redshifts

If we used a perfect set of measuring tools to survey the universe, we would measure its size accurately. But we would miss the essential fact that the universe began at a finite time in the past, and is expanding toward an uncertain, but perhaps not unknowable fate. The discovery of the expanding universe starts with the nature of the atoms that make up the stars.

Stars are made of atoms that are identical to the atoms on Earth, or in the Sun. Those atoms interact with light in distinctive ways that allow us to learn which elements are present in distant objects, and we can also learn whether the stars are moving toward or away from us. The discovery that the galaxies are moving away from us, the realization that the most distant galaxies are receding most rapidly, and the synthesis of these facts into a vision of an expanding universe make up one of the most powerful pictures in modern science.

In the nineteenth century, chemists and physicists knew that different elements have characteristic colors when they were put in a flame, facts still used by the makers of fireworks. The light from a glowing gas can be analyzed more precisely by looking at its spectrum with a spectroscope, which employs a prism or a grating to split the light into colors that make it up. The interesting thing is that while a hot glowing wire, like the filament of a light bulb, gives off light over the whole spectrum from red to blue, a glowing gas does not. Instead it has bright pure colors separated by dark regions where pure hydrogen or pure helium or pure oxygen gives off no light. We now interpret that pattern of bright spectral lines as showing us the internal structure of the atom, but for our purposes it is enough to realize that the spectrum lines of each element provide a fingerprint as distinctive as the whorls and loops on your thumb. We know there is hydrogen in the Sun because the same pattern of spectrum lines shows up in the Sun as in a tube of hydrogen prepared by a chemist on Earth. In fact, the element helium was discovered from its pattern of lines in the Sun before it was isolated on Earth! The atoms in the Sun form their spectra by identical rules, without any variation. Atomic physics has no designated hitter option.

Subtle measurements show that we can detect the motion of objects by observing a shift in the overall pattern of lines from an element. The line has a characteristic color or frequency associated with it. If the glowing object is moving toward us, the lines are shifted toward the blue or toward higher frequency and when it is moving away from us, the lines are shifted to the red and lower frequency. In our usual imaginative way, astronomers call the former a blueshift and the latter a redshift. Either type is called a Doppler shift, after the discoverer of the effect (Figure 2.7).

All light travels at the universal top speed: the speed of light. Although the light emitted by an approaching source travels at the same speed, we see the light shifted to higher frequency. In fact, the amount of the shift can be directly related to the fraction of the speed of light that the source is traveling. This means that in everyday life, where we move slowly compared to 300 000 kilometers per second, these shifts are not noticed by our eyes, although the police use electronic measurements of Doppler shifts in their radar waves to measure the speed of unmuffled motorcycles and baseball scouts use them to measure the zip on a fastball.

We do hear the shift in frequency that motion produces in sound waves. When a fire engine shrieks past with its siren wailing, we can

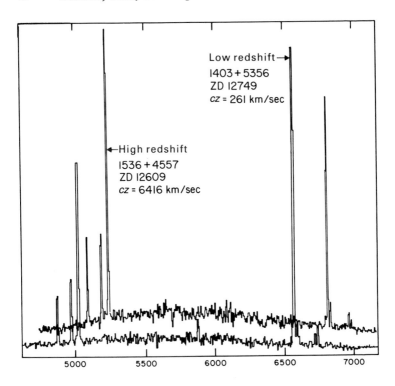

Figure 2.7 Galaxy spectra. Actual spectra obtained at the Smithsonian's Whipple Observatory in Arizona show two galaxies with strong emission lines due to emission from gas. The lower galaxy is at low redshift, the upper at a higher redshift. Note that the pattern of lines is similar, but that the whole set is shifted to longer wavelengths. (Smithsonian Astrophysical Observatory.)

hear the pitch change as the noisemaker goes by. The pitch, which corresponds to frequency, is higher when the fire engine is approaching, and it suddenly drops in pitch as the hook and ladder flashes by. The speed of sound is about 1200 kilometers per hour, so it is not so uncommon to hear shifts of 10 percent in frequency produced by sources, like unmuffled motorcycles, moving at 10 percent of the speed of sound. These are then followed by screeching police sirens to provide a second demonstration.

Perhaps a homely example will help make this connection between frequency and motion more clear. Imagine an itinerant pigeon fancier, Mr. Roller, who goes on a long trip. People are always encouraging pigeon fanciers to go away. Suppose he brings along a flock of trained pigeons, schooled to fly home. When he leaves home he says to his wife, 'I'll send you a note every Monday, my little dove.'

As he travels the first week, his journey takes him far enough away so that the pigeon takes a whole day to bring his billet-doux to his spouse. He launches it on Monday, and the message comes home to roost on Tuesday. During the following weeks, he travels still farther afield, and releases the bird on Monday, as promised, but the pigeon takes an extra day now, since he is farther away, and his tender note takes until Wednesday to reach home. Mrs. Roller begins to worry that Mr. Roller is losing his marbles, sending messages every 8 days, instead of every seven as promised. On the third week, her worst fears are confirmed, since the pigeon takes off on Monday, but due to the added distance Mr. Roller has traveled, it takes until Thursday to get there. Another 8 days between messages!

Then Mr. Roller turns around, and starts to come home. He launches the next pigeon on Monday and, now that he's closer, it arrives on Wednesday. Mrs. Roller is agitated! The previous message came on Thursday, now it's Wednesday – only 6 days have passed. The next week, he emits a pigeon on Monday and she gets it 6 days later on Tuesday. Finally he gets home, and explains that he's been sending out pigeons at the same frequency, but since the pigeons have farther to travel each week while he is receding from home, and less distance each week while he is approaching, the frequency with which the birds arrive is one every 8 days while he is outbound and once every 6 days while he is homeward bound. Mrs. Roller is enlightened, and apologizes for suspecting that he didn't know if he was coming or going.

Now an atom in a star moving away from us emits its light waves at the same frequency it always does, just as Mr. Roller always sent his pigeon out every 7 days, but like Mrs. Roller, we receive them at a lower frequency: shifted to the red.

The application of Doppler shift measurements to astronomy has revealed broad new patterns of motion for objects at different distances. Measurement of velocities for nearby stars show that the Sun's neighbors are milling around gently, with some approaching and some receding. From measurements of more distant stars and gas, we see that the great flat pinwheel of the Milky Way is rotating, one turn every 230 million years. (As described in detail later by Vera Rubin, the rotation of galaxies also provides evidence for the way mass is distributed in galaxies.) Measuring the motions of entire galaxies is the road to understanding the timescale of the universe.

Measuring the spectrum of a galaxy was a challenging problem at the beginning of this century. From 1912 to 1925, Slipher at the Lowell Observatory developed techniques that allowed him to measure velocities for 40 galaxies. There were two surprising results. First, the velocities were very big, up to almost 2000 kilometers per second, much larger than the velocities for stars in our own galaxy. Second, almost all the velocities were redshifts, indicating that the galaxies were moving away from the Milky Way.

Hubble used these data, and his own distance measurements for galaxies, to make a profound discovery. In 1929, he found that there was a simple relation between the distance to a galaxy and its velocity of recession. Working with Milton Humason at Mount Wilson, Hubble obtained redshifts and distances for faint and distant galaxies. This work demonstrated that the distance and the redshift are in direct proportion to one another. If you look at a galaxy twice as far away, it has twice the redshift; three times as far away, three times the redshift. This relation between distance and redshift is known as Hubble's Law and the number that gives the ratio of velocity to distance is called the Hubble Constant.

Today, we have much more efficient ways of measuring redshifts for galaxies than were available to Slipher or to Humason. They used photographic plates as their detectors, but even very good photographic materials only record about 1 percent of the light that falls on them. We use various electronic tools, including solid-state detectors like the charge coupled devices (CCD) that make video cameras work so well in dim light. These thin wafers of silicon are often 50 percent efficient. This means that a 100-inch telescope today collects data 50 times faster than it could in Humason's time. The result is a rapid increase in the number of measured redshifts, from a few hundred in the 1950s to over 25 000 today (Figure 2.8).

Figure 2.8 This typical page of galaxy redshifts is taken from a catalog maintained at the Harvard–Smithsonian Center for Astrophysics by John Huchra. There are now some 25 000 galaxies with measured redshifts. (Harvard–Smithsonian Center for Astrophysics Redshift Survey.)

Name	α (1950)	δ	m_B	V_h	σ	Source	Type	UGC	Comments
0000+0810	00:00:00.0	08:10:00	14.90 1	28663±	42	27	D		A2694
I5378	00:00:00.0	16:22:00	17.00 1	±			15	00001	
0000+4439	00:00:00.0	44:39:00		±				00002	
0000-0357	00:00:03.5	-03:57:00	14.00 7	34560±	58	1731			CO2-31
0000-3612	00:00:06.0	-36:12:50		6474±	30	-1			
0000+2705	00:00:07.5	27:05:00		14881±	33	1731			CO2-29
0000-3618	00:00:09.0	-36:18:08		7690±	30	-1			
0000-3612	00:00:09.6	-36:12:27		15356±	33	1731			CO2-24
0000-4415	00:00:10.6	-44:15:15		14676±	29	1731			CO2-26
0000+1837	00:00:11.0	18:37:00	14.80 1	12100±		31	1B	00003	FAIRALL 626
0000-3626	00:00:12.0	-36:26:16		7847±	32	-1			CO2-16
0000-3611	00:00:13.1	-36:11:43		43111±	65	1731			CO2-37
0000-3621	00:00:15.8	-36:21:22		14215±	34	1731			CO2-20
0000-3622	00:00:18.8	-36:22:13		38975±	211	1731			CO2-25
N7812	00:00:20.7	-34:30:48	14.39 8	38883±	100	3901	2		CO2-10
0000-3626	00:00:21.0	-36:26:03		6845±	30	1731			
0000-3600	00:00:21.2	-36:00:04	16.10 B	42757±	74	3110	-5		
0000-3311	00:00:21.2	-33:11:50	16.40 B	14750±	480	3110	-5		
0000-3613	00:00:21.5	-36:13:53		9421±	44	1731			CO2-21
0000-3604	00:00:24.0	-36:04:34		16085±	39	1731			CO2-22
0000-3623	00:00:24.6	-36:23:22		14450±	44	1731			CO2-15
0000-3616	00:00:27.1	-36:16:57		15508±	52	1731			CO2-44
0000-3614	00:00:28.7	-36:14:47		14452±	33	1731			CO2-45
0000+0356	00:00:29.9	03:56:00	15.50 1	14038±	53	1731			
0000+1836	00:00:30.0	18:36:00	15.50 1	8014±	39	-1	4A	00004	
0000-3616	00:00:30.0	-36:16:09		14969±	45	-1			CO2-49
0000-3608	00:00:30.5	-36:08:24		14549±	30	1731			CO2-40
0000-3613	00:00:31.1	-36:13:10		14685±	31	1731			CO2-50
0000-3611	00:00:33.7	-36:11:10		15436±	34	1731			CO2-47
0000-3607	00:00:33.7	-36:07:38		14083±	37	1731			CO2-47
0000-0211	00:00:36.0	-02:11:00	14.30 1	7323±	31	1731			
0000+2142	00:00:36.0	21:42:00	14.40 1	6605±	32	27	4X	00005	
0000+2454	00:00:36.0	24:54:00	15.50 1	7729±	36	27	5 P	00006	SY2,MK334,IVZW1
0000+3045	00:00:36.0	30:45:00	15.00 1	±		-1			
0000-3614	00:00:38.8	-36:14:46	14.40 8	13378±	44	1731			CO2-11
0000-3612	00:00:39.0	-36:12:53		14936±	30	3901	-2		
0000-3612	00:00:39.7	-36:12:54		14900±	39	1731			CO2-43
N7814	00:00:41.1	15:52:03	11.71	1050±	8	2812	2A S	00008	
0000+0320	00:00:42.0	03:20:00	15.30 1	8009±	31	-1			

The velocities we observe are not small. It is easy to determine redshifts for galaxies receding at 30 000 kilometers per second, 1/10 the speed of light, and, with great effort, galaxies that are moving away from us at 2/3 the speed of light can be measured. Since the distance is proportional to the velocity, these rapidly receding galaxies are very distant, and the light from them has been traveling to us for most of the age of the universe. In this sense, a telescope is a time machine that allows us to glimpse things as they were, when the universe was young. The most distant objects for which redshifts have been measured are not galaxies, but quasars, which are widely thought to be exceedingly powerful, compact sources of light and radio emission at the centers of some galaxies. The object with the highest redshift to date emitted its light when the universe was only 20 percent of its current age. The most distant thing we could ever hope to see would be an object whose light has been traveling toward us since the Big Bang.

The next step in our quest for measuring redshifts is to equip telescopes on the ground with many, perhaps as many as 100, small fiber optic feeds that bring the light from 100 different galaxies into a spectrograph for analysis. Then our present telescopes will allow us to measure redshifts 100 times more rapidly, and we can move from custom-made, one-at-a-time, redshift craftsmanship to the mass production of redshifts. For the most distant galaxies, which are also the faintest, the construction of very large telescopes, such as the 10-meter Keck telescope now being built at Mauna Kea is the way to gather enough light to measure their redshifts, and to understand their natures.

For many galaxies, the detailed distance estimates from Cepheids or other reliable methods are not available. In those cases, it is not unusual to assume that the Hubble Law works, and to derive a distance from the redshift alone. The technique that has been used so successfully to map the locations of galaxies in three dimensions has been to use a measured redshift and the Hubble Law to determine the distance to an individual galaxy. This method is the one I used with Augustus Oemler, Stephen Shectman, and Paul Schechter to map the Bootes Void, and it is the method used by Margaret Geller and her collaborators to map the large-scale structure of the universe as described in the next chapter.

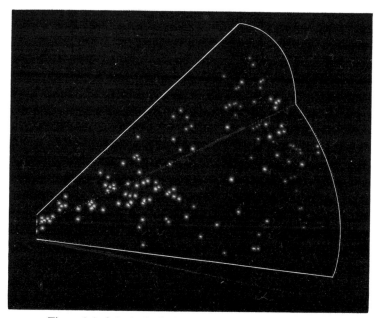

Figure 2.9 Galaxies in the Bootes Void sample. The sample studied by Kirshner, Oemler, Schechter, and Shectman is depicted here in side view. In this view, the Earth is at the apex of the triangle, and the distance to each galaxy, determined by redshift measurement, is shown. The scale of the diagram is such that the galaxies at the extreme right are about 1 billion light years from Earth. Note the large vacant region near the right side of the diagram. This empty area, about 27 million cubic light years in volume, is the Bootes Void, the largest known 'desert of galaxies.' (Wolbach Image Processing Laboratory/Harvard–Smithsonian Center for Astrophysics.)

In the case of the Bootes Void (Figure 2.9), our measurements showed that a large patch of the sky, where we expected to see many galaxies, had nearby galaxies at a redshift of 10 000 kilometers per second and very distant galaxies at a redshift of 20 000 kilometers per second, but none with redshifts over a large interval from 12 000 to 18 000 kilometers per second. This vast empty region, a desert in the forest of galaxies, we called the Bootes Void. It was the largest feature known in the distribution of galaxies, and it is a clue that the fabric of the universe is woven in a more complex way than we have yet mapped.

We can use Hubble's Law as a distance tool without thinking about it, but it is much more stimulating to consider the meaning of Hubble's Law. Does the recession of galaxies from us signal a kind of cosmic halitosis, with the other galaxies rushing away from us? A view like that is not only a sign of paranoia, it assumes that we occupy a special place, a unique place in the scheme of the universe. If there is one lesson from our slowly unfolding tale of cosmic distances it is that our location is not unique and our view is not special. We are not at the center of the solar system, the Sun is not in the center of our galaxy, and our galaxy is just one among uncounted billions of galaxies. Why then should we assume that we are at the center of the universe? It might be more prudent to adopt the opposite view: that our view must be the typical one.

No matter how nice an argument sounds, it has to agree with the facts. At first glance, the facts seem to show that all the galaxies are rushing away from us. Is there some reasonable way to reconcile that observation with the idea that our view is typical? Surprisingly, the answer is yes.

Imagine a giant jungle gym, with galaxies at every intersection. Suppose you are invested with the power to make that jungle gym grow in all directions. (Maybe it's made of living bamboo!) What would that look like, from the point of view of any individual galaxy? If the jungle gym doubled in size, then a galaxy's closest neighbor would be twice as far away – at a distance we might call 2 JU (for Jungle Units) – and it would be receding. The most distant neighbors, let's say ten junctions away, will also double in distance, ending up 20 JU away. So, as viewed by that particular galaxy, all the others are receding, and the more distant ones are receding faster. In fact, an expanding framework produces Hubble's Law, with all galaxies seen as moving away at velocities of recession just proportional to their distances.

Now here's the interesting part. We looked at the expanding jungle gym from the point of view of one galaxy. But we could have picked any galaxy, and we would have gotten the same picture, the same Hubble's Law. If you're on a galaxy moving away from me, you think I'm moving away from you. This very simple picture, in which all of space is expanding, gives a Hubble's Law picture for every observer. It covers the facts as we see them, it does not require that we have a special position in the universe and it has become the working model for modern astronomy. Its basis in observation springs from the measurement of distance, and winds through the rugged path of parallax and standard candles to connect those measurements with the redshifts of the galaxies around us.

One interesting consequence of this picture is that it leads to an estimate for the age of the universe. If the nearby galaxies are receding slowly, and the distant galaxies receding rapidly, you can imagine a past for the universe, in which any galaxy you care to pick out was close to us. If you calculate the time in the past when that was true, it turns out to be the same, whether you are considering a nearby galaxy or a distant one. One way to think about this is to say that the expansion began a finite time ago. If we know velocities (from redshifts) and distances (from some other method), then we can calculate the age of the universe. The uncertainties in the distance scale come back to haunt us: the age of the universe calculated this way is between 10 and 20 billion years. Since the oldest stars are in the same age range, we are close to having a reliable estimate of the time when the universe began. The measurement of redshifts and of distances leads us all the way back to the deepest questions about the origin of the universe.

Further reading

Berendzen, R., Hart, R., and Seeley, D. *Man Discovers the Galaxies*, New York: Science History Publications (Neale Watson Academic Publications), 1976.

Ferris, T. *The Red Limit* (2nd Edition), New York: Quill, 1983.

3

Mapping the universe: slices and bubbles

MARGARET J. GELLER
Harvard–Smithsonian Center for Astrophysics

For more than 4500 years human beings have been making maps. We have mapped the surface of the Earth in almost excruciating detail. Indeed, the concept of map making extends to regimes we cannot explore directly – we make maps of the positions of atoms in solid materials and in DNA, the key to life itself. From the smallest to the largest physical systems, maps reflect the human drive to know. Today, one of the grand frontiers is the exploration of the universe over distances of hundreds of millions – even billions – of light years.

Even a cursory review of the history of maps of the Earth† shows the strong ties between maps and our understanding of the physical world. One of the earliest maps (Figure 3.1) shows the small, round, flat world of the Babylonians in the sixth century BC. Water completely surrounds the Tigris–Euphrates valley. The seven small circles around the periphery of the land mass are the mythical mountains of Biblical fame.

Even some 2000 years later, maps were still often influenced more by myth and superstition than by measurement and science. The Hereford map made at the end of the thirteenth century (Figure 3.2) covers a significantly larger fraction of the Earth than the Babylonian picture, but it still bears a remarkable resemblance to its ancient ancestor. The flat, round Earth is surrounded by water. The land masses are distorted to fit the preconceived pattern and to place Jerusalem at the center of the map. England (the frontier) is at the

---

† The major reference for this history of maps of the Earth is *The Mapmakers* by J. N. Wilford, New York: Vintage Books, 1982.

Figure 3.1 Babylonian map *ca* 500 BC. (Courtesy of the Trustees of the British Museum.)

lower left. The Spanish and Italian peninsulae are only barely recognizable jutting into the oversized Mediterranean.

The age of exploration beginning in the fifteenth century brought a general recognition of the importance of maps as scaled representations of the surface of the Earth. Accurate maps became a matter of life and death. The sixteenth century map in Figure 3.3 is actually derived from a second century Ptolemaic map. The map is still distorted in many ways. Ptolemy realized that features on the Earth could be placed according to their longitude and latitude, but these coordinates were very poorly known even for the portions of the globe which had been explored. For example, the features in this map extend over too many degrees of longitude; they are all stretched along the East–West direction. Nonetheless, this map clearly represents an advance toward an accurate representation of the structures on the surface of the Earth.

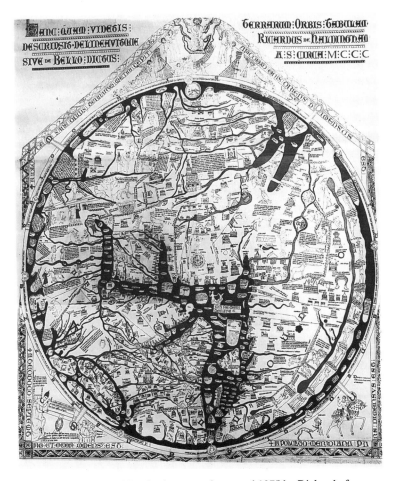

Figure 3.2 Hereford map made around 1275 by Richard of Haldingham. (By permission of the Dean and Chapter of Hereford Cathedral.)

Ptolemy's profound understanding of the purpose of map-making was essentially lost for thirteen centuries. 'Geography is a representation in pictures of the whole known world together with the phenomena contained therein,' he wrote, adding that the goal of map-making is 'to survey the whole in its just proportions.' (As quoted in *The Mapmakers*, pages 27–8.) In other words, a map is a scaled model of the salient features of a physical system. This perspective makes the connection between science and map-making

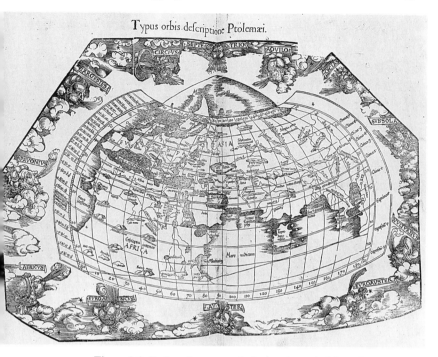

Figure 3.3 Ptolemaic map made during the early fifteenth century. (By permission of the Houghton Library, Harvard University.)

obvious. Science is also a process of making models – a process of controlled abstraction.

After the discovery of the New World, the art of map-making advanced rapidly. In spite of the increasing sophistication of cartography and the accelerated pace of discovery, the outlines of the continents (except for Antarctica) were first defined reasonably well only in the early nineteenth century. As late as 1801, Matthew Flinders first established that Australia is a continent, not a group of islands.

During the early twentieth century, clear knowledge of the large-scale geographic features of the Earth (particularly the shapes of the continents) led to a revolutionary idea. Before 1912, Alfred Wegener noticed that the continents can be fitted together like pieces of a jigsaw puzzle and made the creative leap to suggest they once were a single giant land mass which later fragmented. He further argued that the continents had drifted apart to their present locations. The idea of a dynamic Earth was greeted with more than a bit of skepticism – it was even ridiculed. Today, plate tectonics, which explains continental drift, is a full-fledged physical theory with abundant observational support. Some of the most convincing evidence has accumulated only over the last 30 years as we have become able to explore the ocean floor. In 1974, the rate of spreading of the ocean floor was measured directly by a team of French scientists who traveled to the floor of the Atlantic rift valley in a bathysphere. They measured a rate of $2\frac{1}{2}$ centimeters per year, stunningly close to the predictions of the theory. Extraordinary expeditions like this one are among the tools of the scientist-mapmakers who continue to piece together an explanation of the origin and evolution of the Earth's surface structure.

Continental drift is but one example of the connection between maps and science. Another well-known example is the way that knowledge of the structure of DNA has opened the way to understanding the basis of life. A third is the current attempt to map the universe in order to gain an understanding of the distribution of galaxies over scales of hundreds of millions of light years. Knowledge of the formation and evolution of these very large structures is closely tied to an understanding of the universe as a whole.

How do we go about exploring the structure – making a map – of a system as dauntingly large as the universe itself? What does it mean to make a map of the universe? The volume of the visible universe is of the order of 10 000 billion billion billion (10^{31}) cubic light years! When we talk about making a map of some region of the universe, we really mean that we plot out the distribution of galaxies like our own Milky Way. In the visible portion of the universe there are billions of such bright galaxies. (See the chapter by Vera Rubin for a description of the properties of galaxies.)

Practical limitations dictate that a feasible study of the distribution of galaxies could include measurements for 10 000 – or perhaps 100 000 galaxies. Current surveys include a few thousand galaxies. In other words, we have mapped out a fraction of the universe comparable with that fraction of the Earth covered by the state of Rhode Island – about 1/100 000.

Although we are limited to mapping out a small fraction of the volume of the universe, we would like to try to answer a well-defined, perhaps deceptively simple, question: how big are the structures – the patterns (if any) – in the distribution of galaxies? Once we focus on the question, the challenge is to sample the universe in such a way that we obtain the best possible limits on the properties of big structures (if they exist).

An imaginary voyage to a cloud-enshrouded Earth-like planet provides a lesson for surveyors of the universe. The planet may have large geographic features. You set out to find out about these features, *but* you can only map out 1/100 000 of the surface area of the planet. The single region you map can have any shape. What shape should you choose? How about a patch? Figures 3.4 and 3.5 show two Landsat photographs of the Earth. Each one covers a bit more than 1/100 000 of the Earth's surface. The first image (Figure 3.4) of a small

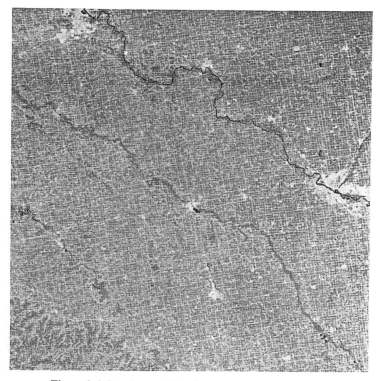

Figure 3.4 Landsat satellite photograph of a portion of the Great Plains of the United States. (NASA photograph.)

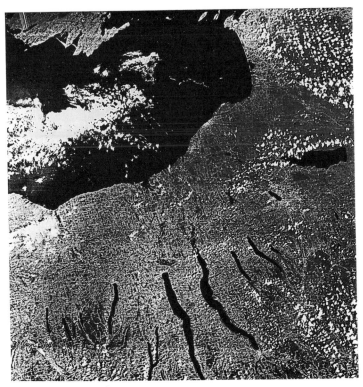

Figure 3.5 Landsat photograph of the Finger Lakes region of New York State. (NASA photograph.)

patch of the great Corn Belt of the upper mid-Western United States might lead you to think the Earth is essentially featureless; the second image (Figure 3.5) of the Finger Lakes region in New York State indicates that the Earth has lots of land, pock-marked with smaller bodies of water. Wrong again! Three-quarters of the surface of the Earth is covered with water. Thus, a randomly selected small patch of the surface of the Earth would be just a patch of ocean three-quarters of the time. One such patch would also lead to the erroneous conclusion that the Earth is completely covered with water. In contrast, imagine a sample which is a long thin strip spanning the circumference of the Earth. Most strips like this will cross both continents and oceans and will yield some measure of the size of both features. We can apply this lesson directly to making maps of the

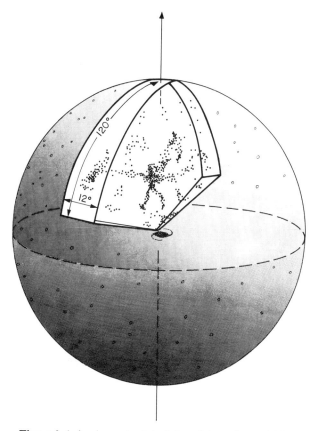

Figure 3.6 A schematic of the 'slice of the universe.' The galaxy is an oversized representation of our own Milky Way where the Sun is well away from the center. The diameter of the Milky Way is about 75 000 light years. We look out into the wedge-shaped portion (white region) of the universe where we see the arrangement of galaxies (the 'human figure'). The outer edge of this wedge is 450 million light years from Earth. (Center for Astrophysics illustration by John Hamwey.)

positions of galaxies in the universe. Because this map will be a three-dimensional one, the analogy to a strip is a wedge (Figure 3.6).

The first step in making a map is the construction of a catalog of the positions of galaxies on the sky. Fortunately, during the 1960s, Fritz Zwicky and his collaborators catalogued the positions of more than 30 000 galaxies, most of them nearer to us than about 300 million light years. The catalog is based on the then-new collection of Palomar Sky Survey plates of the northern sky. This collection of photographic plates was taken during the 1950s with the 48-inch Schmidt telescope. Each of the plates covers about 36 square degrees of the northern sky. This survey has been an invaluable guide to the positions of stars in our own galaxy and to the positions of galaxies external to our own. Many galaxies more distant than those in Zwicky's catalog are also clearly visible on these photographic plates.

Figure 3.7 shows the distribution of galaxies in one portion of the sky surveyed by Zwicky. The coordinates are celestial coordinates; right ascension is the celestial longitude and declination is the celestial latitude. Each point in the plot represents a galaxy. The 7035 galaxies in this plot are not uniformly sprinkled across the sky – there are patches of low density and there is an obvious dense condensation of galaxies at 13 hours and 28°, the Coma Cluster. The universe contains many clusters of galaxies like Coma; each of these contains hundreds to thousands of galaxies bound together by their mutual gravitational pull. The galaxies in a cluster orbit one another at velocities of the order of 1000 kilometers per second; yet, even at this velocity, it takes a billion years for galaxies to travel the several million light years across the cluster.

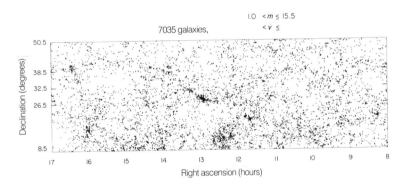

Figure 3.7 Distribution of galaxies on a portion of the sky. Right ascension is the celestial longitude; declination is the latitude. Each of the 7035 points is a galaxy; most are within three hundred million light years of Earth. The ticks at 26.5, 32.5 and 38.5 degrees show the declination limits of the two complete survey strips which stretch across the entire right ascension range in this plot. (Center for Astrophysics illustration from the Zwicky Catalog.)

Although the distribution of galaxies in the two-dimensional map (Figure 3.7) isn't uniform, it is not exactly a striking pattern. However, as we are about to see, the three-dimensional map is something to write home about!

Until the 1930s extragalactic astronomers were limited to the 'flatland' of maps like Figure 3.7. In 1929 Hubble showed that the universe is dynamic – one of the most profound revelations of modern science. The universe expands in such a way that each galaxy recedes from us with a velocity proportional to its distance from us. If we can find a way to measure the velocities, we can make a three-dimensional map.

We measure velocities of galaxies by examining their spectrum, that is, the amount of light they emit as a function of the color, or wavelength, of the light. In each spectrum there are narrow features (an excess or deficit of light over a narrow range in wavelength) known as spectral lines. Each of these lines is the signature of particular atoms or ions in the galaxy we observe. For example, in many galaxies we see the lines of sodium, the same element responsible for the yellow glare of street lights and for the yellow light you see if you throw salt into a gas flame. But the sodium lines in the spectrum of a galaxy are at a longer (redder) wavelength than the lines of low-pressure sodium street lights. The lines are *redshifted*.

For Hubble, measuring a redshift was enormously time-consuming. It took a whole night to measure the redshift of a single galaxy. Thus, a project to map even a small portion of the universe was well beyond the limits of the existing technology. Today, astronomical instruments take advantage of the enormous recent advances in solid state detector technology. These instruments have a much higher efficiency for the detection of photons than traditional photographic plates. The advent of this technology has brought an explosion in the amount of data available to chart the continents and oceans of the universe. It is now possible to use a modest-sized telescope to do work which used to be the province of only the largest instruments.

The Smithsonian Astrophysical Observatory operates a 1.5-meter telescope (small by today's standards) located on Mount Hopkins, just south of Tucson, Arizona. Every clear, moonless night, John Huchra and I use this telescope to measure redshifts for galaxies within about 300 million light years of the Earth. The redshifted photons which we detect have been traveling unimpeded through space for hundreds of millions of years before they hit the mirror of our telescope poking up from this speck called Earth.

To measure a typical redshift in our survey, we collect photons from each galaxy for about half an hour. Within the next 5–10 years (depending on the weather in Tucson!), we plan to measure redshifts for some 12 000 galaxies. During the spring of 1985, Valérie de Lapparent, Huchra, and I completed the first sample in this project. We measured redshifts for approximately 1100 galaxies in a strip 6° wide stretching 120° across the sky. The strip extends across the entire right ascension range in Figure 3.7, with a declination range of 26.5–32.5°. Why this particular strip? The number of galaxies included is about what we can measure in one spring of observing time on the 1.5-meter telescope. The strip is roughly overhead in Tucson and seemed a convenient starting place. Most important we chose a geometry for the sample which is sensitive to large patterns in the distribution of galaxies – patterns which persist over distances of 100 million light years or more.

Once we have measured the redshifts we have three coordinates: right ascension, declination, and redshift. We can then plot the positions of galaxies in a map like the one in Figure 3.8. In this map we are at the vertex of the cone. The map shows only two of the three dimensions of the survey; the declination coordinate is perpendicular to the page. (See Figure 3.6 for a schematic of the three-dimensional geometry.) In the azimuthal direction, we plot the right ascension – the coordinate on the sky which runs along the long dimension of the strip. Along the radial direction, we plot the velocity derived from the redshift. Each point is a galaxy comparable with the Milky Way. Perhaps the map has a subliminal appeal because the pattern of galaxies looks like a human figure. In any case, the pattern is striking!

We designed a survey to look for big structures – and we found some. Nearly all the galaxies lie in thin, sharp structures which loop across the map. In general, the galaxies toward the outer (curved) edge of the map are farther away from us; the distance to them is just proportional to the velocity we plot. The constant of proportionality, the Hubble Constant, remains uncertain by about a factor of 2. A velocity of 10 000 kilometers per second corresponds to a distance between 300 and 600 million light years (I will use the smaller estimate

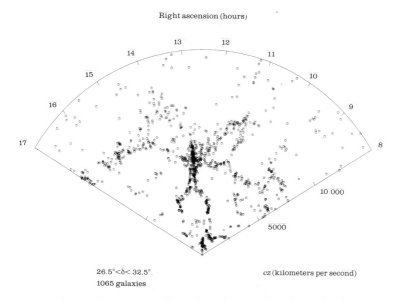

Right ascension (hours)

26.5°<δ< 32.5°
1065 galaxies

cz (kilometers per second)

Figure 3.8 Map of a slice of the universe. We are located at the point
at the bottom of the map. Along the radius we plot *cz*, the redshift,
or velocity of recession. The long dimension of the strip, the right
ascension or celestial longitude, is plotted along the azimuthal
direction. The latitude direction is perpendicular to the page. The
range of latitude runs from 26.5° to 32.5°. Each of the 1065 points is
a galaxy roughly equivalent to the Milky Way. Note the vast voids
(empty regions) and the thin structures looping across the map. The
Coma Cluster is the 'torso' of the 'human figure.' (Center for
Astrophysics illustration.)

throughout this chapter). If you treat the pattern of galaxies as a
connect-the-dots exercise, your pencil outlines the vast dark regions
we call voids. These regions contain few if any bright galaxies. The
largest of these voids (centered at 15 hours and 7500 kilometers per
second) has a diameter of 5000 kilometers per second, or 150 million
light years. At velocities greater than about 10 000 kilometers per
second, the density of galaxies in the maps decreases because, at these
redshifts, we see only the few intrinsically brightest objects.

The map of recession velocities is not exactly a map of distances.
There are distortions because of the motions of galaxies internal to
systems like the Coma Cluster, the torso of the 'human figure'. The
cluster is stretched out along a line pointing toward us at the origin.

The length of this 'finger' is a measure of the typical motions of galaxies within the cluster. From the width and length of this finger, we can estimate the amount of matter in the cluster by applying Newton's laws. If we measure the amount of light from the galaxies in the cluster and figure out the mass of stars responsible for the light, the total falls far short of the estimate based on Newton's laws. We come up short by a factor of 10! In other words, we see only about 10 percent of the mass; the rest is dark. In these large systems the missing mass or dark matter problem is even more extreme than for individual galaxies. (See the chapter by Vera Rubin.)

Other distortions of the velocity map occur if all the galaxies in some region have a motion relative to the expansion of the universe. We call these motions large-scale flows. In general, we expect these velocities to be small compared with the extent of the voids in Figure 3.8. Except for the fingers associated with rich clusters, we expect that the big patterns in our velocity map are a reasonable representation of the patterns we would see if we could measure distances directly. From here on, we treat the map for the most part as though the 'velocity' label reads 'distance'.

Although the map in Figure 3.8 is only a thin slice, we can use it to figure out the most likely three-dimensional arrangement of bright galaxies. We have two guides in searching for the answer to this puzzle: pure geometric intuition and theoretical models for the way the distribution of galaxies might look.

Let's examine a couple of models in order to get a feeling for the power of the data to discriminate among them. Extragalactic astronomers have long recognized that galaxies are not uniformly distributed in space. There are many clusters of galaxies similar to the Coma Cluster. For some time, it seemed that clusters of galaxies might be more or less randomly distributed in a sparse sea of other galaxies. The sharp structures in the map immediately rule out this possibility.

In the early 1970s, the late Soviet scientist, Yaakov Zel'dovich suggested a model to explain the distribution of galaxies on large scales. The colorful (and informative) name of this model is the 'pancake' picture in which matter is initially distributed over enormous sheets which fragment to form galaxies and clusters of galaxies. Computer simulations based on these ideas produce a three-dimensional web-like structure in which most galaxies are strung out along thin, thread-like structures, and with clusters of galaxies forming at the intersections of the threads. The threads are, in turn, the locations of the intersections of the original 'pancakes' or sheets. On the basis of this model, other Soviet scientists like Jaan Einasto were among the first to argue that large voids in the distribution of galaxies must be common. However, unless the slice in Figure 3.8 is somehow special, the pattern produced in this model differs from the data. In general, few filaments of a web would lie in a thin slice through it. If we were slicing through a web we would expect most of the galaxies in the slice to be in separated clumps, not in the connected network we see. In other words, if our slice is typical, the thin structures where the galaxies lie are not one-dimensional threads or strings.

What sort of geometric structure would yield the pattern we see in almost any slice? There are several familiar things which fit the bill. Imagine, for example, taking a slice through the soapsuds in your kitchen sink. Make the slice thin compared with the diameter of a typical bubble. The surfaces of the bubbles – the films of soap – would outline a pattern very much like the one defined by the galaxies. The interiors of the bubbles correspond to the voids. Of course, a soap bubble is about 27 orders of magnitude smaller than a void in the galaxy distribution and the physics of soapsuds is very different from the physics of large-scale structure in the universe. Only the geometries are similar. Other familiar structures like honeycombs and sponges also have similar geometries, but the distinctions among these are subtle and currently beyond the discriminatory power of the data. We do know that the distribution of galaxies is made up of thin sheets in which the galaxies lie. These sheets surround or nearly surround vast voids. The sheets which appear to run all the way across the survey are the surfaces of several adjacent 'bubbles'.

Our generalizations about large patterns in the distribution of galaxies are based on a map of less than 1/100 000 of the visible universe. Obviously, there is no shortage of *chutzpah* in extragalactic astronomy. Although we have been careful about the design of the surveys, our result may not be so different from a model of the Earth's surface derived from an individual Landsat image. As we map out larger and larger regions of the universe, we are likely to revise our picture continually.

Since the completion of the first slice, we have finished surveying

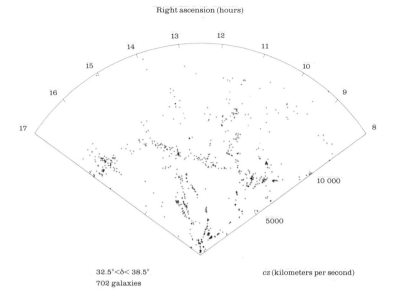

Figure 3.9 A second slice of the universe adjacent (to the north) to the one in Figure 3.8. The declination range for this slice is 32.5–38.5°. There are 702 galaxies. Are these structures the natural extensions of the ones in Figure 3.8? (Center for Astrophysics illustration.)

three more slices. Slices adjacent to the one in Figure 3.8 provide a test of our hypothesis about the topology of the structures. If the pattern is a result of a slice through thin sheets, we should see their natural extensions in the adjacent slice. Figure 3.9 shows the pattern of galaxies in the slice just to the north of the first. Once again the slice on the sky extends from 8 hours to 17 hours and has a width of 6° in declination (32.5–38.5° as marked in Figure 3.7). The 702 galaxies in this new slice are again in thin structures.

Does the pattern in the new slice meet our expectations based on the first one? Figure 3.10 shows the data for the two slices together. The circles are the galaxies in the first slice; the pluses the galaxies in the second. The correspondence is remarkable. The structures in this combined slice are thicker than those in either of the slices taken separately because the structures are curved or tilted with respect to the plane of the slice.

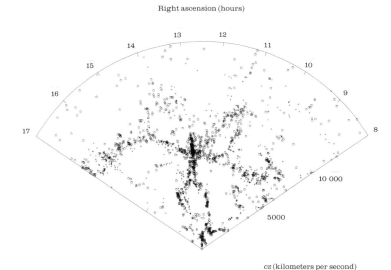

Figure 3.10 The 12° strip which includes the two slices shown in Figures 3.8 and 3.9. The galaxies from Figure 3.8 are circles; the ones from Figure 3.9 are pluses. Note the remarkable correspondence of the structures in the two slices. The structures are thicker here because they are tilted or curved with respect to the slice. (Center for Astrophysics illustration.)

Surveys of other regions reveal similar structures. Robert Kirshner and his collaborators have explored the void in Bootes, a distant empty region comparable in scale with the largest holes in Figure 3.8. Riccardo Giovanelli and Martha Haynes have used the Arecibo radio telescope to measure redshifts for galaxies in the Perseus–Pisces region. Here, too, there is evidence for connected structures and voids. In the southern hemisphere, Luis da Costa and his collaborators have just completed a shallower survey again indicative of the kind of structure we see in Figures 3.8, 3.9, and 3.10.

The pattern we observe in the distribution of galaxies has profound implications. One of the simplest and perhaps most disconcerting is that the largest voids we see are the largest we can detect within the limits set by the dimensions of our survey. There are, at the moment, no direct limits on even larger structures. These enormous dark, but not necessarily empty, regions are common. Every survey big enough to contain one does so. These results make it clear that if we could sample the universe in volumes 100 million light years on a side, the distribution of galaxies would differ substantially from one volume to the next. Some volumes ('patches') might be devoid of bright galaxies; others might contain lots of galaxies in one or several thin sheet-like structures.

How big a region do we have to survey in order that one region looks essentially the same as another? The diameters of the largest voids we have seen so far are about 1 percent of the radius of the visible universe – still small. If we find similarly striking voids with much larger diameters we might have to abandon one of the fundamental ideas in cosmology, the idea that the universe is homogeneous on some sufficiently large scale. We have clearly not obtained a complete answer to our question about the extent of large structures, but we have made progress. We have found a well-defined pattern and the next step to look for even larger structures is evident. We have to make larger surveys which reach to greater distances.

The development of the large voids and sheets in the universe depends upon many poorly known or even unknown properties of the universe and its contents. Perhaps the most basic of the unresolved issues is the overall average density of matter in the universe, for this factor determines both the geometry and the history of the universe. If the density is large enough, the universe is gravitationally bound, or closed. In other words, the universe has a finite extent and a finite lifetime: it will expand to some maximum radius and then collapse to a hot, dense end. Conversely, if the density is too low, the universe is infinite and we say that it is unbound, or open: it will expand forever and will become increasingly dilute. The 'critical density' value is one that marks the transition between the open and closed models.

Although the elegant inflationary model of the early universe (see the chapter by Alan Guth) is based on arguments for a critical density, there is no convincing direct observational evidence for or against this value. On the other hand, the motions of stars in galaxies and of galaxies in clusters support an average density only between 1/10 and 2/10 of the critical value, favoring an open universe. Even at this lower value we are faced with a serious problem: 90 percent of the matter in the universe is invisible. At the critical density, 99 percent of the matter must be dark.

The physical properties of this dark material are issues important for understanding large-scale patterns in the distribution of galaxies. Particle physicists who study the fundamental structure of matter have recently offered several elementary particles as candidates for the stuff which dominates the large-scale dynamics of the universe. These particles are predicted by a variety of theories, but so far none of the cosmologically interesting ones have been detected in the laboratory. (In the next chapter, Vera Rubin will describe in detail the exhaustive efforts to determine both the distribution and nature of dark matter.)

The action of gravity on large scales is an integral component of any explanation of the formation of structure in the universe. Galaxies and clusters of galaxies are held or 'bound' together by the force of gravity, but these large systems probably developed from smaller concentrations of matter in the early universe. The origin and properties of these small lumps and bumps are still puzzles. However, we think that as the universe evolves, small irregularities in the distribution of matter tend to grow because the gravitational attraction of a particular lump overcomes the cosmological expansion in the region near it. The increasing mass and density of the lump affect the motions of matter in larger and larger surrounding regions; the lump continues to grow as more and more surrounding matter collapses onto it. At the same time that gravity causes clustering, it also causes low-density voids to expand and to pile up matter around their edges. Qualitatively, gravity acts to increase the lumpiness of matter in the universe.

But does gravity alone produce large-scale clustering of galaxies over tens or hundreds of millions of light years? In spite of the many uncertainties in the models, computer simulations show that without a very special arrangement of matter at early times, gravity is inadequate to the task. In the simplest models, the contrast between high- and low-density regions is not large enough, the voids are too small, and the remarkable coherent patterns in the galaxy distribution are absent.

The challenge to the standard models can be understood from a simple physical argument. If the positions of galaxies tell us where matter in the universe is now, then the big empty regions must have been made by moving matter out of these regions during the lifetime of the universe. In that time – 10–20 billion years – matter moving at the typical velocity of galaxies today, some 300 kilometers per second, travels only 10–20 million light years, a distance small compared with the 100-million-light-year-diameter scale of the voids observed.

One way out of this dilemma is to appeal to another puzzle – the dark matter problem. Because we do not see the dark stuff, we don't know what it is or where it is. If we could look directly at the way all matter, rather that just light-emitting matter, is distributed in the universe, we might see that it is actually very uniform and that galaxies form preferentially where the density of matter is somewhat higher than average. Because higher density regions are rare, the galaxy formation process automatically leaves large holes without having to move galaxies across enormous distances. Although computer simulations based on these ideas produce results which resemble the data, it is worrisome that we still do not know what process determines where galaxies form.

Moreover, if these models are right, the striking contrast between the voids and sheets in our maps may be misleading. Galaxies could be mere flotsam and jetsam or a uniform sea of dark matter. Although the phenomenology of galaxies might be interesting in itself, they would be poor guides to the dynamics of the universe as a whole. The dark voids would be full of stuff – dark matter and perhaps faint galaxies which would not be included in our surveys.

There are other ways to remedy the inadequacy of the simplest gravitational models. Jeremiah Ostriker and Lennox Cowie, and independently Satoru Ikeuchi, suggested that explosions occurred when the universe was less than a billion years old and drove the hydrogen into thin shells which fragmented to form galaxies. This suggestion is appealing because it is easy to imagine how it produces a bubble-like pattern. The evolution of the shells of gas formed by such early explosions is complex. The largest structures we see could be the result of a merging of several adjacent shells. (Here, too, it may be that the galaxies are not good tracers of the way matter in the universe is distributed. For example, if the shock wave from an explosion swept up ordinary matter and left much less interactive elementary particles behind, the voids could again be full of dark stuff.) But these models also have their drawbacks. Although there are many explosive, energetic phenomena in the universe, ranging from supernovae to the sources of energy in the nuclei of galaxies, there are none which release sufficient energy over a short enough period to make the largest observed shells. Even worse, if there were such enormous explosions we probably should have detected their effect as irregularities in the very smooth microwave radiation background.

Because of the complexity of the physics involved in the formation of galaxies and larger structures, it is hard to use the current epoch we observe to guess what happened at much earlier times in the universe. Many theories offer plausible explanations of the qualitative features of the observed structure. However, the sheer number of physical processes dictating how the early, small 'seeds' might evolve is so large, that the models tell markedly different stories. For example, in some models galaxies are the first large structures to form. Gravity then causes them to cluster together to make the larger patterns we see in our surveys. In other models, the large structures form first, as enormous sheets, or 'pancakes,' then fragment to form galaxies. Because the age of the universe is long compared to the timescale for the formation of individual galaxies, these strikingly different pictures can both lead to reasonable approximations to the universe we observe nearby today. Perhaps only direct observations of a more 'ancient' universe at high redshift will ultimately reveal whether galaxies or larger structures form first.

We have made maps of a portion of the nearby universe. It should also be possible, in principle, to make maps of the arrangement of galaxies in the universe at earlier epochs. As we look out in space we look back in time. Light from more distant objects takes a longer time to reach us. It takes 300 million years for light to reach us from an object 300 million light years away, a billion years for light from an object a billion light years away, and some 10–20 billion years – the age of the universe – for the photons in the microwave background. We see the universe, or objects in the universe as they were 300 million or even billions of years ago. For example, we see the most distant galaxies in the maps in Figures 3.8–3.10 as they were 450 million years ago. Although this time is long by human standards, it is short compared with the time it takes for the structures we see to change. By mapping out the distribution of galaxies at much larger redshifts we *can* see how the structures looked in the past. Just as we use the geologic record to study the evolution of the Earth's crust, we can use observations of distant objects to see how they and the universe have evolved.

Unfortunately, our current 'maps' of the distant universe at 'high redshift,' cover only very small regions and are generally two-dimensional. For example, the surveys by David Koo and Alex Szaley and by Tony Tyson, John Jarvis, and Frank Valdes reach back about 70 percent of the age of the universe, or 7–15 billion years, but these surveys cover only a minute fraction of the sky – comparable to the fraction covered by the Moon. (A region of this size may contain thousands of fuzzy images, each about 1/1000 of the angular diameter of the Moon but each a full-fledged galaxy.) These surveys can provide only the *position* of the galaxies on the sky, because most of the objects are too faint for the measurement of redshifts, even with the largest telescopes. The distances to the faintest galaxies can only be guessed by comparing their brightness with that of galaxies nearby.

Still, we can learn about the way galaxies cluster by examining their two-dimensional distribution on the sky, as we can see by looking again at Figure 3.7. Indeed, the distribution of galaxies in the deep surveys can be compared with the distribution in the shallow surveys as a guide to the evolution of the structure. For instance, Koo and Szaley have found that the clustering of galaxies in their deep survey is remarkably similar to clustering of galaxies nearby, but they caution that redshift measurements are necessary before reaching any firm conclusions.

Maps of the 3 degrees Kelvin radiation background may provide yet another clue to the origin of the structure we see. This microwave background, discovered by Arno Penzias and Robert Wilson in 1965, is often called 'the whisper of the Big Bang,' and its existence indicates that the very early universe was hot as well as dense.

When the universe was younger than about a million years, matter was hot enough to be ionized. In this state, radiation was tied – coupled – to the matter. When the temperature of the universe dropped to about 4000 degrees Kelvin, the electrons combined with nuclei to form neutral atoms. At this point, the radiation could no longer be scattered effectively by the matter and we say that radiation and matter 'decoupled.' Any lumps in the matter distribution at the time of decoupling should reveal their presence as irregularities in the radiation background. More precisely, we would expect small variations in the temperature of the radiation background seen today to correspond to small fluctuations in the distribution of matter at these very early times. Sensitive searches for these temperature fluctuations have so far yielded only upper limits. On a scale of 1.5 arcminutes, Juan Uson and Dave Wilkinson find that the fluctuations in temperature are less than a few parts in 100 000. Although upper limits seem less interesting than detections, these measurements indicate that matter distribution in the early universe was remarkably uniform. Indeed, the extraordinary smoothness of the radiation background is a strong constraint on theories for the formation of galaxies and larger structures and may completely rule out some models.

The explosive model for the formation of galaxies and larger structures may be one of those most severely limited. In such a model, the energy of any explosion strong enough to produce a large void, is also sufficient to reionize matter and to rescatter the microwave background photons. In this case, the microwave background would provide a sketch of the universe at a time between ten and a few hundred million years after the Big Bang, a time when the explosions were occurring.

The microwave background photons are the oldest ones we can use to observe the properties of the early universe directly. At epochs when the matter is ionized, the universe is opaque because photons cannot travel far without being scattered from their path. If we finally detect structure in the background radiation, we will probably make an enormous advance in understanding galaxy formation, but the puzzle of the origin of the lumps in the primordial matter distribution will remain. This puzzle may be solved by scientists working in the world of the very small – the world of particle physics.

Advances in particle physics may even lead to the identification of the major constituents of the mysterious dark matter which appears to dominate the large-scale dynamics of the universe. Laboratory experiments are already underway to look for particles which go by the colorful name 'wimps,' or weakly interacting massive particles. Among these particles, the best dark matter candidate is the axion. Although they may be the primary constituents of the universe, none has ever been seen. Detection of one of these particles would be a revolution in both particle physics and cosmology.

The unanswered questions in cosmology are profound. I often feel that we are missing some fundamental element in our attempts to understand the large-scale structure of the universe. It is both sobering and exciting to think again about the analogy between mapping the Earth and mapping the universe. It took centuries to learn the shapes of continents well enough to prompt the intuitive suggestion of continental drift. Then, it took another half-century to substantiate the theory.

Perhaps our approach to mapping the universe is more sophisticated than the early approaches to mapping the Earth. But the universe is immense; we are encumbered because we can only observe it from afar. The techniques required to uncover the fundamental physics governing the largest realm of nature may be well beyond our current abilities.

Mapping the universe will undoubtedly keep us busy, awed, and fascinated for a long time to come. I often ask myself what we will learn about large-scale structure during my lifetime. There will be surprises, answers to old questions, and the uncovering of new puzzles. At every stage we will think we understand, but at every stage there will be nagging doubts in the minds of those who wonder.

Further reading

Bartusiak, M. *Thursday's Universe*, New York: Times Books, 1986.

Ferris, T. *Galaxies*, New York: Stewart, Tabori, and Chang, 1982.

Gribbin, J. *In Search of the Big Bang*, New York: Bantam Books, 1986.

Hubble, E. *The Realm of the Nebulae*, New York: Dover, 1958.

Pagels, H. *Perfect Symmetry*, New York: Bantam Books, 1986.

Silk, J. *The Big Bang*, San Francisco: W. H. Freeman, 1980.

Wagoner, R. and Goldsmith, D. *Cosmic Horizons*, San Francisco: W. H. Freeman, 1982.

Weinberg, S. *The First Three Minutes*, New York: Basic Books, 1977.

Wilford, J. N. *The Mapmakers*, New York: Vintage Books, 1982.

4

Weighing the universe: dark matter and missing mass

VERA C. RUBIN
Carnegie Institution of Washington

Introduction

Does the distribution of light in the universe map the distribution of matter? Until the present decade, astronomers rarely asked this question, but most would have answered in the affirmative. Recent observations, however, have produced compelling evidence that at least 90 percent of the matter in the universe, and perhaps as much as 99 percent, is not radiating at any wavelength. This dark matter, invisible in the optical, ultraviolet, X-ray, gamma-ray, infrared, and radio regions of the spectrum, is detected by its gravitational attraction on the matter which we can see.

As long as 50 years ago, there was initial evidence that something was amiss in our understanding of the motions of stars in our galaxy, and the motions of galaxies in clusters. Due principally to a total lack of comprehension of these observations, but also to a lack of large telescopes and suitably sensitive detectors, astronomers generally ignored these early findings. During the past decade, however, observations of galaxies in the optical, the radio, and the X-ray spectral regions have convinced astronomers of the presence of the ubiquitous dark matter. Astronomers no longer call this missing mass as they once did, for it is the light, not the matter, which is missing.

As our concept of the universe has undergone this major and rapid change, the manner in which we study the universe has also been altered. Thirty years ago, observational cosmology consisted of the search for two numbers: H_0, the rate of expansion of the universe at the position of the galaxy; and q_0, the rate of deceleration of the expansion. Twenty years ago, the discovery of the relic radiation from the Big Bang produced another number, 3 degrees Kelvin. But it is the past decade which has seen the enormous development in both observational and theoretical cosmology. The universe is known to be immeasurably richer and more varied than we had thought. There is growing acceptance of a universe in which most of the matter is not luminous.

The most compelling evidence for the existence of dark matter comes from studies of velocities of stars and gas as they orbit the centers of their galaxies. Astronomers had long believed, by analogy with the solar system, that orbital velocities would be highest for stars near the center, while orbital velocities would be slowest for stars far from the center. (In the solar system, where virtually all of the matter is located in the Sun, Mercury, the closest planet, orbits with a velocity about ten times as great as that of Pluto, the most distant planet). However, detailed measurements of velocities of stars and gas in several hundred galaxies show that stars at the periphery of a galaxy orbit with speeds as great or greater than those close to the nucleus. The conclusion is inescapable: the stellar orbits remain high in response to the gravitational attraction of matter which we cannot see.

Nature has played a trick on astronomers, for we thought we were studying the universe. We now know that we were studying only the small fraction of it that is luminous. In order to appreciate this altered view of the universe, I would like to start by describing the matter that we can see, and how it has taught us about the dark matter. Among the most important players in this drama are the spiral galaxies, so I will start by describing what we know about these objects.

Spiral galaxies and the distribution of matter

When you view the sky on a moonless night from a dark site in the northern hemisphere, every star which you see with the naked eye is a member of our own galaxy. This means that these stars are all bound to each other by the dominant force in the universe, gravity, and that they are orbiting in concert about a common, distant center. The stars are not distributed at random, but most are concentrated to a flattened disk, a disk in which the Sun (our own star) resides. When we, living on our Earth and orbiting our Sun, look out into space along this plane, we see a band of stars projected onto the sky; this band we call the Milky Way. A classical problem of antiquity was to learn the composition of the Milky Way. It was not until 1609 that Galileo turned his newly perfected telescope to the Milky Way and resolved the glowing band into individual stars. 'Do you know what the Milky Way is made of? I do,' he says in Bertold Brecht's play *Galileo*. Indeed, he had taken the first step toward deducing the structure of our galaxy.

The plane of the Milky Way is home to many objects besides single stars; it is composed of double stars, clusters of stars, tenuous clouds of atomic hydrogen, and dense clouds of molecules containing hydrogen, oxygen, carbon, and other elements, as well as particles of dust and metallic grains. These components, many of them opaque to optical light, cloud our view of the distant regions of our galaxy. Thus, to learn more of the details of the distribution and motions of stars in a galaxy, we must turn our telescopes on galaxies beyond our own, especially those in directions away from the disk of the Milky Way.

Stars in a typical galaxy are arranged in two major components: a disk and a bulge. We expect that our galaxy, when viewed edge-on, would appear much like NGC 891 (Figure 4.1(a)). In this view, the clouds of gas and dust dim the light from the billions of stars in the disk, producing the mottled appearance. Light from the other billions of individual stars in the bulge, the second dominant component, is blurred into one intense glowing central region. Although it is difficult to photograph our galaxy so as to show its resemblance to external galaxies, it is not impossible. Figure 4.1(b) is an all-sky photograph taken from Chile when the center of our galaxy was overhead. The dense star clouds, mottled by the obscuring gas and dust, show a galaxy closely resembling those distant galaxies we observe with our telescopes.

Properties of 'disk stars' differ dramatically from those of 'bulge stars.' The brightest stars in the disk are hot, young, massive, and blue, having newly condensed from the gas contained in the disk. Their circular orbital motions reflect the circular motions of the gas from which they formed. One orbital period for a star located half-way out in the disk (comparable to the location of the Sun in our galaxy) is 200 million years, about 1/100 of the age of the universe. But many massive stars will not live to complete one orbit about the center of their galaxy. They will be born, grow to maturity, evolve rapidly, and die before one orbital period has elapsed.

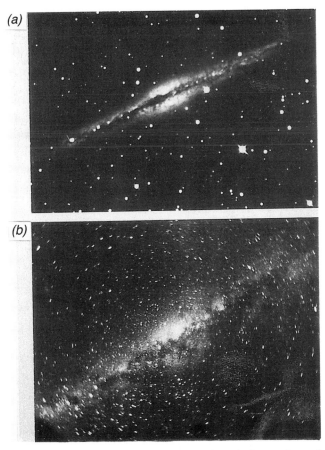

Figure 4.1 (*a*) NGC 891, a spiral galaxy viewed edge-on, located at a distance of 40 million light years. Note the prominent equatorial lane of gas and dust which partially obscures the radiation from the stars in the disk. (Hale Observatories photograph.) (*b*) A view through the plane of our Milky Way, looking toward the galactic center. The photograph was taken from the CTIO when the center of our galaxy was overhead. Note the remarkable resemblance of our galaxy to NGC 891. The distance to the center of our galaxy is about 20 000 light years. (Photograph © Association of Universities for Research in Astronomy, Inc.)

Figure 4.2 A region of the Gum Nebula, a star-forming region in our galaxy visible in the southern sky, showing a dust globule with a bright rim. The nebula contains small, dense, molecular clouds, some of which are still forming stars. The Yale 1-meter telescope at CTIO was used with a Carnegie RCA C33063 image intensifier, plus Hα filter. (Photograph courtesy of Dr. J. A. Graham.)

In contrast, the stars in a bulge are cool, old, of small mass, and red. They have been evolving very slowly ever since their birth early in the lifetime of the galaxy; their lifetimes are enormously long. Their orbital motions carry them on elongated orbits far out into the halo of the galaxy and then back near the center. These orbits reflect the eccentric motions of the gas clouds early in their evolutionary history. At present, the lack of gas in the bulge precludes the formation of new stars there. (Not all galaxies have both bulge and disk components like our own: some are virtually all bulge, some are all disk.)

It takes two things to form a star: gas, and something to compress the gas to a high density. In our galaxy, appropriate conditions exist only in the disk. A region of high density, once established, gravitationally attracts matter to itself, forming a protostar. Nearby atoms tumble in and heat up the protostar's core. This process takes place not in full view, but deep within a dense molecular cloud; most of the action is hidden from us. In the Gum Nebula (Figure 4.2), a region of active star formation relatively near our Sun, the dark clouds are silhouetted by the light from the embedded young stars.

Ultimately, the temperature at the center of this protostar reaches millions of degrees, and nuclear processing of matter begins. A star has been born. During the millions or billions of years of the star's lifetime, the star radiates by transforming hydrogen and helium into heavier metals deep in its interior. During this process, some stars continually shed matter into the region between the stars. We know some of these objects as planetary nebulae (Figure 4.3). But, ultimately, the light elements in the core are depleted; and, with the star's energy source gone, a true 'energy crisis' ensues. The central core collapses and the outer gaseous layers of the star are regurgitated into the interstellar regions as the star explodes as a supernova (Figure 4.4). This gas between the stars, now richer in the heavier elements produced in the star, mingles with the nearby gas clouds. Future generations of stars born from this brew will be composed of a larger fraction of heavy elements than were the initial generations. In a very real sense, during its lifetime, a star is a factory for making all of the matter in the universe heavier than hydrogen and helium. All of the other elements which went into forming our solar system, forming our planet, forming our bodies, and forming the ink and paper which makes it possible to print these words, were produced deep in the

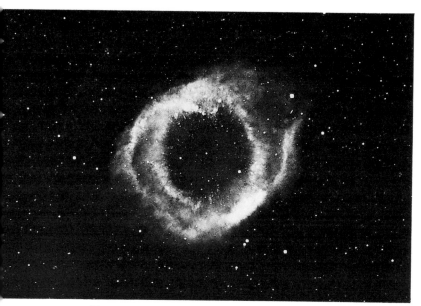

Figure 4.3 Planetary nebula, NGC 7293, in the Constellation Aquarius. (National Optical Astronomy Observatories photograph.)

Figure 4.4 Supernova in Centaurus A discovered in 1986. (European Southern Observatory photograph.)

interior of some star. (With an insight truly profound, Jon Muir wrote in his 1911 book *My First Summer in the Sierra*: 'When you try to pick out anything by itself, you find it hitched to everything else in the universe.' What a great observer he was.)

Galaxies rarely exist alone, and our galaxy is no exception. If you live in the southern hemisphere, two satellite galaxies of the Milky Way, the Magellanic Clouds, are a familiar sight in the night sky. (If you were lucky enough to have viewed the southern sky in spring 1987, you would have seen in the Large Magellanic Cloud the first naked-eye supernova since the time of Kepler and Tycho Brahe in 1604. See Figure 4.5.) The Magellanic Clouds are small galaxies with irregular structure, much gas, and uncounted regions of active star formation. We know that these objects are galaxies because they contain the usual galaxy population of stars, star clusters, hydrogen gas, molecules, and supernovae remnants. Astronomers used to believe that the Magellanic Clouds were young galaxies, for their gas content and their present rate of forming stars is enormously high. But we now know that they contain an underlying population of old stars, stars as old as those in our bulge. It seems some remarkable circumstance has recently (in an astronomical sense) initiated star formation in the Magellanic Clouds. Most likely, this was a passage of the Clouds through the disk of our galaxy.

The Magellanic Clouds are gravitationally bound to our galaxy, and orbit the center on eccentric paths which carry them through our disk, like other halo population objects. Presently, we can only guess at the form of their orbits, but in 10 years or so, astronomers will be able to measure the very tiny displacements of stars in the Magellanic Clouds from their positions recorded on earlier photographic plates. However, we have been able to make some knowledgeable guesses about their paths through our galaxy, because they have a left a very telltale signature.

In a normal galaxy, the relative separations of stars are very large. If two galaxies pass through each other at high speed, the stars will pass each other generally unaffected, although some stars from the smaller object will be lost to the more massive one. However, the gas will be significantly disturbed. Indeed, an enormous tidal tail of hydrogen emanating from the Magellanic Clouds is assumed to be a relic of an earlier passage through the disk of our galaxy. The future of

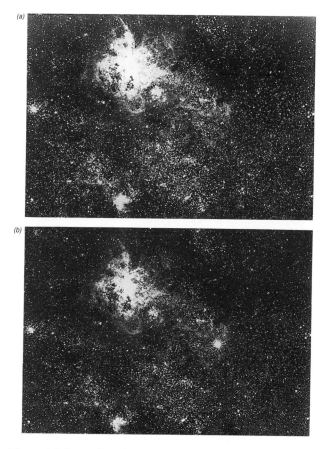

Figure 4.5 Large Magellanic Cloud (*a*) before and (*b*) after SN 1987A. (National Optical Astronomy Observatories photograph.)

the Magellanic Clouds is bleak, as it is for most small galaxies living in the environs of a large galaxy. Successive passages through the plane of our galaxy will probably fragment the Clouds still more. Ultimately, they should lose their independent identities and merge with our galaxy. This circumstance indicates how critically a galaxy is a product not only of its heredity, but also of its environment.

At distances greater than the Magellanic Clouds, our galactic region is populated by about 20 galaxies, which astronomers, with

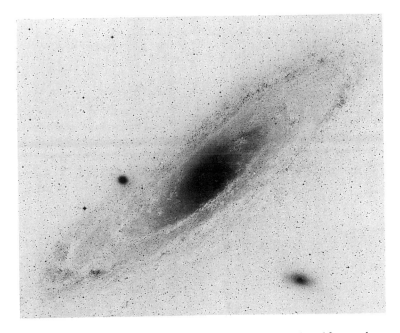

Figure 4.6 The Andromeda Nebula, M31, reproduced from a plate taken at the Palomar 48-inch Schmidt telescope, and kindly made available by Dr. Morton S. Roberts.

little imagination, call the Local Group. The spiral galaxy (Figure 4.6) seen beyond the constellation of Andromeda (named M31, the entry number 31 in the catalogue produced by Messier in 1784) is the only other large massive spiral in the Local Group. It, too, has several small satellite galaxies which orbit its center. The remaining members of the Local Group are generally small, low-mass, irregular aggregates, with little resemblance to the majestic spirals. The Local Group, in turn, is an outlying member of a large agglomeration of several thousand galaxies, many forming into subclustered units, which we call the Virgo Supercluster (Figure 4.7).

Presently, one of the most exciting areas of extragalactic astronomy is the study of the distribution and motions of matter on the largest scales. Stars form into galaxies, galaxies into clusters, and clusters into rich superclusters. Understanding the large-scale distribution and motions of the rich superclusters will tell us much about the early history of the universe. At present, we are literally in the dark; we do

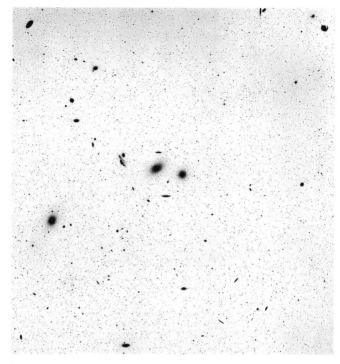

Figure 4.7 A region near the center of the Virgo Cluster, from a Palomar 48-inch Schmidt plate. The giant elliptical galaxy M87 is at left below center. (Hale Observatories photograph.)

not know if galaxies formed and then aggregated into clusters, or if clusters formed, later to fragment into galaxies. On all scales which we can observe, the distribution of clusters and superclusters is very lumpy. Nowhere yet do we observe the smooth, isotropic, homogeneous universe we learned about in school. A plot of the million brightest galaxies (Figure 4.8) shows stringy lace-like regions surrounding vast regions void of bright galaxies. The work of Geller, Huchra, and de Lapparent described elsewhere in this book shows how the patterns of sheets and voids grow larger on ever larger scales. This distribution imposes very tight constraints on the composition and distribution of matter in the early universe. Any acceptable description of galaxy formation and evolution must reproduce this lumpy distribution existing today.

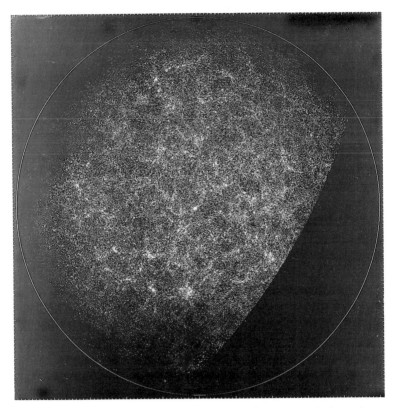

Figure 4.8 An all-sky plot of the million brightest galaxies as they appear on the northern sky, from counts made by Shane and Wirtanen at the Lick Observatory. The Coma and Virgo Clusters of galaxies are located near the center of the image. Note the striking lacework pattern and the conspicuous voids. (Figure courtesy M. Selder, B. L. Siebers, E. J. Groth, and P. J. E. Peebles and *The Astronomical Journal*; © 1977 The American Astronomical Society.)

Dark matter in spiral galaxies

This then is the universe that we see with our eyes and with our telescopes, starting out at our galaxy with its bright bulge and flattened disk, and extending at present to the rich superclusters. But for astronomers and physicists who wish to do more than just catalog the distribution of galaxies in the universe, progress comes by applying two fundamental guidelines: (1) the laws describing the behavior of matter in our laboratories on Earth are the same laws which describe the behavior of matter in distant galaxies; and, (2) gravity, which acts in a simple, predictable manner, is the dominant force in the universe. Newton's simple relation which describes the fall of an apple to the Earth, and the fall (i.e., orbiting) of the Moon about the Earth and the Earth about the Sun, is equally correct in describing the orbiting of a star about the center of a galaxy. Thus, one way to study the distribution of matter in a galaxy is to study the motions of stars as they orbit the center of the galaxy.

For most of my professional life, I and my colleagues have been attempting to learn about the distribution of mass in the universe by studying the distribution of mass in spiral galaxies. Before I describe our observations, it may be helpful to review how celestial objects respond to the gravitational force acting on them and how that response can reveal the large-scale distribution of matter.

By the end of the seventeenth century, Robert Hooke had suggested that the planets were subjected to a gravitational force from the Sun, a force whose intensity decreased as the square of the distance. Newton then recognized that all pairs of objects in the universe have a gravitational attraction for each other that is proportional to the product of their masses and inversely proportional to the square of the distance between them. If the distance between two objects is increased by a factor of 2, their mutual attraction is decreased by a factor of 4.

In our solar system, virtually all of the mass is contained within the Sun. Planets orbit the Sun because of the gravitational attraction arising from the Sun's large mass. Planets are continually falling toward the Sun, but their forward motion insures that they continually 'miss' hitting it. Within the solar system, orbital speeds of the planets decrease with increasing distance from the Sun. Mercury, the closest planet to the Sun, orbits with a velocity of 47 kilometers per second; Pluto, 100 times more distant, orbits with a speed 1/10 as large, 4.7 kilometers per second (Figure 4.9). Any object at Pluto's distance will orbit with this same velocity, be it comet, rock, or planet. As I will discuss in more detail below, this is what Galileo was attempting to demonstrate when he dropped two balls of different masses from the tower at Pisa.

If we know the mass of the Sun, we can use Newton's Law of Gravitation to predict the orbital velocity for an object at any distance

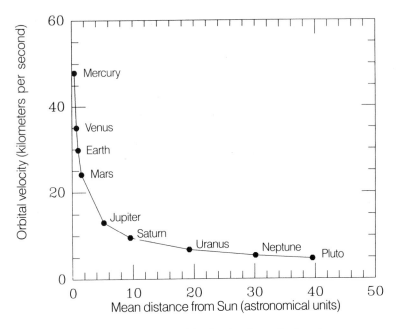

Figure 4.9 Orbital velocities for the planets in the solar system, plotted as a function of their mean distances from the Sun. The decrease in velocity with increasing distance, accurately described by a $1/r^{1/2}$ curve, is evidence that virtually all of the mass in the solar system is located in the Sun.

from it. It is a remarkable fact of nature that we can equally well turn the problem around. If we know the distances and the velocities of the planets, then we can determine the mass, i.e., the amount of matter, within the Sun. And, by extension, if we can determine the velocities and the distances of stars as they orbit a galaxy, we can determine the amount of matter interior to each stellar orbit.

In a galaxy, the brightness is strongly peaked near the center, and falls off rapidly with distance. Astronomers had long inferred that the distribution of mass was indicated by the distribution of luminosity. Understandably, astronomers believed that the density, too, was strongly peaked to the center and decreased rapidly with distance. Then, by analogy with the solar system, they expected that orbital velocities of stars would be very small at the faint limits of a galaxy. Thus it was a surprise to learn that in all galaxies studied, orbital velocities remain high, even at the optical edge of the galaxy where there is almost no light. This result follows from observations carried out by my colleagues and myself, among others, during the past decade; I describe these below.

Astronomers can measure motion by noting the change of position of an object on the sky. Only for the closest stars in our own galaxy is it possible to detect such changes. For more distant objects, the change in position is too slight to be observed. Even for our near neighbor, the Andromeda Galaxy, it would take some 2000 years for an orbital velocity of 200 kilometers per second (a velocity comparable to the Sun's orbital velocity) to carry a star 1/10 of a second of arc across the sky. An angular separation this small can be detected optically from the Earth only with great skill and sophistication. To study motions in distant objects, a different technique is needed – one based on the phenomenon of the Doppler shift.

If the disk of a spiral galaxy is oriented so that its plane is sharply tilted with respect to the line of sight from the Earth, the orbital motions on one side of the nucleus will carry the stars and gas in that galaxy toward our galaxy, and those on the other side of the nucleus away from our galaxy. Just as the pitch of a train whistle changes as it approaches and recedes from the listener, so the characteristic frequencies emitted by the atoms of gas in a galaxy will be increased or decreased as the atoms approach or recede from us. We use a spectrograph attached to a large telescope to observe the light emitted by the hydrogen atom in many locations across a distant galaxy; the displacement of the spectral emission line from its stationary (laboratory) position is a measure of the velocity of that gas toward or away from the observer (Figure 4.10). Due to this Doppler shift, the spectral lines of the approaching material will be blueshifted, or raised in frequency; the lines of the receding material will be redshifted, or lowered in frequency. An accurate measurement at any point on a spectral line will reveal the velocity along the line-of-sight corresponding to the point on the galaxy where that radiation was emitted.

Only during the past decade or two has it been possible to get high-resolution optical spectra of the faint outer regions of galaxies. The present availability of large optical telescopes, of high-resolution, long-slit spectrographs, and of efficient electronic imaging devices have made these observations feasible. Over 10 years ago my col-

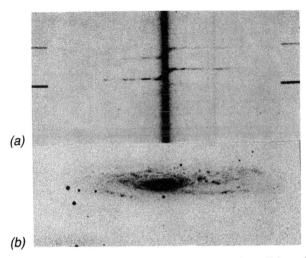

(a)

(b)

Figure 4.10 (*a*) Emission lines of Hα and [NII] from disk gas in NGC 801. The strong vertical continuum arises from stars in the nucleus. As a result of the rotation, gas on the right is receding from the observer, so the emission is shifted toward the red; gas on the left is approaching the observer, so the emission is shifted toward the blue. This photographic spectrum was taken with the KPNO 4-meter telescope plus spectrograph plus image tube, on a baked and preflashed IIIa-J plate. (*b*) NGC 801, a spiral galaxy seen close to edge-on, from a plate taken at the KPNO 4-meter telescope by Dr. B. Carney.

leagues and I set out to measure the rotational velocities completely across the luminous disks of suitably tilted spiral galaxies. Our aim was to study the internal dynamics and distribution of mass in individual galaxies as a function of their morphological (bulge-to-disk ratio, size) properties. Most of the observations have been obtained with the 4-meter telescopes at the Kitt Peak National Observatory (KPNO) in Arizona and at the CTIO in Chile. A few of the spectra were recorded with the 5-meter telescope at Palomar, and the 2.5-meter telescope at the Las Campanas Observatory of the Carnegie Institution in Chile (Figure 4.11).

Initially, all of our observing was done with image tubes and photographic plates. We had to arrive at the observatory at least 24 hours before the scheduled start of the observations in order to have sufficient time to prepare the equipment. Because we would be

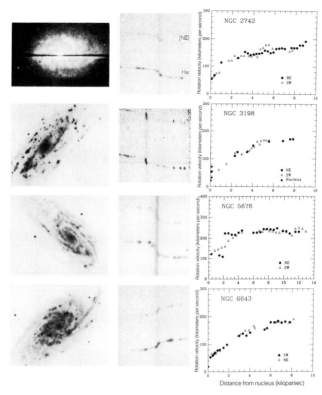

Figure 4.11 Images, spectrograms, and rotation curves for four spiral galaxies. The spectra, obtained at the KPNO 4-meter telescope, are reproduced with wavelength increasing from bottom to top. The vertical line on each spectrum arises from the integrated emission from the stars in the nucleus. The strongest (lower) line on each spectrum comes from hydrogen; above it is a nitrogen line. The spectra were obtained using a CCD. The right panel displays the measured rotational velocities as function of nuclear distance. Note that for all galaxies, velocities remain high at the limits of the optical image. (Figure courtesy *The Astrophysical Journal*; © 1987 The American Astronomical Society. Reproduction of NGC 5676 and NGC 6643 courtesy Dr A. Sandage.)

attempting to record light from the faint outer regions of galaxies, exposure times were long, sometimes up to 3 hours. Thus it was necessary to employ the most sensitive equipment then available. In total darkness, I would cut the photographic plates (glass, not film) into 2-inch squares, place them in an oven and bake them for hours in an atmosphere of dry nitrogen to increase their sensitivity. It was imperative that no specks of dust contaminate the plates; every astronomer has known the agony of exposing for 3 hours on a clear night, only to find that a dust mote has obliterated the very spot of interest. Following the observations, the plates were developed and dried, carried home, and measured under a microscope to give the locations of spectral lines to an accuracy of 1 micrometer.

Today, observations are recorded not on photographic plates, but on digital detectors whose output is stored on magnetic tape. Thus, rather than waiting to scan a wet plate when it is removed from the water some time after the 3-hour exposure, we can now display the image on a computer terminal immediately following its 20-minute integration. And, with some special devices, it is even possible to watch on the terminal as the image is acquired. Measuring, too, is simplified – instead of an eye to a microscope it is a finger on a keyboard – and the job is completed in a fraction of the time previously required.

We have now obtained and measured spectra of over 100 spirals; other optical and radio astronomers have observed probably twice this number, so there is a large body of work available for analysis. From our measurements, we determine the orbital velocity as a function of distance from the center of the galaxy; we make great efforts to measure the velocity out to the edge of the optical disk. One startling result emerges from all of these observations. Virtually all the rotation velocities remain high at all positions far out in the galaxy (Figure 4.12). There are no excessive regions where the velocities fall off with distance from the center, as would be expected if the mass were centrally concentrated. The conclusion is inescapable: matter, unlike luminosity, is not concentrated near the center of spiral galaxies. In short, the distribution of light in a galaxy is not at all a guide to the distribution of matter.

Using conventional Newtonian gravitational theory and a simple model, we can demonstrate the basis for this result with a single

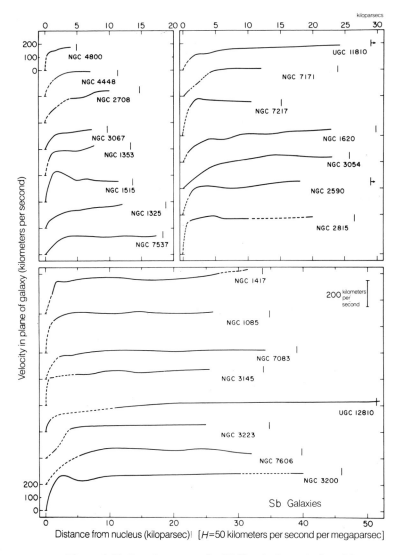

Figure 4.12 Rotation curves for 23 Sb galaxies, galaxies with well-established spiral structure and moderate bulges. Note that virtually all of the rotation velocities remain high at the limit of the optical galaxy. (Figure courtesy *The Astrophysical Journal*; © 1982 The American Astronomical Society.)

equation. Assume that in a disk galaxy, the gas and stars all rotate about the center in circular orbits, with velocities which result from the combined gravitational attraction of all matter in the galaxy. (This is an appropriate model except near the nucleus, especially if there is a prominent bulge component to the galaxy. We neglect a small factor which depends on the geometrical form of the mass distribution.) If the galaxy is in equilibrium, then on any particle of mass m orbiting with velocity V at distance R from the center, the gravitational force is exactly balanced by the centrifugal force. We can then equate the expressions for the gravitational and the centrifugal forces,

$$\frac{GMm}{R^2} = \frac{mV^2}{R} \tag{1}$$

where M is all the mass, luminous plus dark, interior to R. Note that the mass of the orbiting particle, m, can be cancelled from the equation because it appears in both terms. Hence any object, be it a star, a cluster, or even a galaxy, will orbit with the same velocity at the same distance from an interior mass. It is this fact which made Galileo's objects fall with the same velocity. If we adopt units such that G, the gravitational constant is equal to 1, we then can write a second equation:

$$V = \frac{M^{1/2}}{R^{1/2}} \tag{2}$$

In two domains of interest, the solution of the second equation is especially simple and revealing.

(1) Central mass

In the solar system, essentially all of the mass is in the Sun. Thus, the mass M does not increase with distance beyond the Sun. Hence, from Equation (2), the velocity V decreases as $1/R^{1/2}$. Distant planets orbit more slowly, as already discussed. Also from Equation (2), we can predict the orbital velocities of planets, knowing the mass of the Sun. Or, as mentioned above, we can turn the problem around: knowing the velocities of the planets, we can calculate the mass in the Sun. Because the orbital velocities of the planets follow the $1/R^{1/2}$ relation to a very high accuracy (Figure 4.9), we know that virtually all of the mass is located at the Sun. Conversely, for spiral galaxies, the high velocities at large distances R are convincing evidence that most of the mass is *not* restricted to the nuclear regions.

(2) Constant velocity

When the velocity V has a constant value at all R beyond the nucleus, M/R must be constant. Hence the mass interior to R increases linearly with R. This variation describes the pattern we observe in spiral galaxies. The amount of matter interior to the radius R continues to grow linearly as R increases; the mass has not reached a limiting value at the limit of the optical luminosity. Such a mass distribution is very different from the light distribution in a galaxy.

Detailed analyses of velocity curves, based on the above formulae, offer convincing evidence that the stellar orbital velocities remain high in response to the gravitational field of matter which we cannot see. While this dark matter is only a small fraction of the galaxy mass at small distances from the center, it becomes a relatively larger fraction at larger and larger nuclear distances. This is because the discrepancy between the optical luminosity and the calculated mass becomes larger and larger at greater and greater distances from the center of the galaxy. Astronomers generally refer to this unseen matter as a 'dark halo,' although it is difficult to describe the geometry of something which we cannot see and which can only be inferred from its gravitational attraction on luminous matter.

In some galaxies, neutral hydrogen gas is distributed well beyond the optical galaxy, and offers the opportunity to study the dynamics and the mass distribution beyond the limits of the optical galaxy. These 21-centimeter observations show that this gas, too, is orbiting with velocities which remain virtually constant at large radial distances. The 21-centimeter rotation curve obtained by Roberts in 1975 for the Andromeda Galaxy (Figure 4.13), an important milestone in revealing that rotation curves are flat, was followed by numerous other rotation curves. The neutral hydrogen gas is itself not the dark halo, for it amounts to only a small fraction of the total galaxy mass. But its motions act as a probe of the galaxy potential, and thus extend the study of galaxy mass beyond the range which we can see optically. More recent and more sensitive 21-centimeter observations have

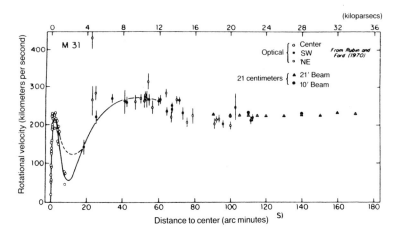

Figure 4.13 Rotation velocities for M31, the Andromeda Galaxy, from optical observations of Rubin and Ford, and from radio observations of Roberts and colleagues.

determined rotation velocities extending to several times the optical diameter for a few galaxies. In all cases, the velocities remain virtually constant beyond the optical image.

The above evidence convinces us that a component of dark matter is clumped around spiral galaxies, with the fraction of dark-to-luminous matter becoming larger with increasing distance from the nucleus. Detailed analyses of specialized observations suggests that at least in some galaxies, the dark matter is distributed in a spherical geometry.

Zwicky (along with Jan Oort, who studied the motions of stars in our own galaxy) deserves credit for being the first astronomer to uncover evidence for the existence of nonluminous matter in external galaxies. As early as 1933, a sufficient number of radial velocities were available for galaxies in the Coma Cluster to allow an analysis of cluster dynamics. Zwicky noted that the individual galaxies were moving so rapidly that their mutual gravitational attraction, calculated from their luminous mass, was insufficient to hold the cluster together. If clusters were not flying apart (and the evidence suggested they were not), then dark matter must be present; its gravitational attraction bound the cluster together.

Initially, astronomers were prone to think that this 'missing mass' was an exotic property of clusters, unrelated to more isolated galaxies. The importance of the recent observational work is that it demonstrates that nonluminous matter is a property of single galaxies as well.

Dark matter in elliptical galaxies

Just as gas and stars orbiting a galaxy can be used as test particles of the galaxy's gravitational field, galaxies in clusters can be used as test particles of the cluster's potential. In addition, the hot X-ray gas surrounding certain cluster ellipticals has been used to determine the gravitational potential and mass distribution for a few such galaxies.

For example, the elliptical galaxy M87, the second brightest galaxy in the Virgo Cluster, has two unique properties: it is located approximately at the center of the cluster, and it is virtually at rest with respect to the mean cluster motion. Observations show that M87 is enveloped by an enormous hot plasma which radiates X rays. This X-ray gas extends at least as far as 1.5° from M87, a distance which encompasses several other Virgo galaxies. Due to its high kinetic temperature, the gas would escape from the galaxy unless there was sufficient mass in the galaxy to retain this corona gravitationally. Daniel Fabricant and Paul Gorenstein have calculated that a mass for M87 equivalent to 50 trillion times that of the Sun is necessary to bind the hot gas.

Thus, for M87, too, the distribution of mass is not revealed by its distribution of light. Astronomers often describe masses, M, luminosities, L, and the ratio of the two, M/L, in units of the Sun's mass and luminosity. Hence, a star with $M/L = 1$ is a star that has one unit of mass for every unit of luminosity. For an agglomeration of stars such as those in a galaxy, a mean value of $M/L = 1$ indicates that the average mass-to-light ratio of the composite stellar population resembles that of the Sun. At distances of 300 000 light years from the nucleus of M87, a mass-to-light ratio of several hundred is implied by the X-ray observations. This means the quantity of dark matter required is so large that there are hundreds of units of mass for each unit of luminosity. Observations such as these confirm that the importance of dark matter increases beyond the limits of the optical galaxy. We do not yet know how far halos extend. The answer to this question will settle a major uncertainty in observational cosmology. Interestingly enough, many years ago Gerard de Vaucouleurs (and later Halton Arp and Francesco Bertola) produced evidence from optical images of M87 suggesting that the galaxy had a faint envelope of light extending several degrees. This optical detection, then just at the limit of the photographic study, was not widely believed.

M87 is the only elliptical galaxy in the Virgo Cluster to exhibit such an extended massive X-ray halo. It is clear that the gravity at the position of M87 is enormous, and may represent the potential of the cluster as a whole. It is also possible that the M87 halo has been acquired from other galaxies in the cluster. Regardless of the evolutionary history, this observation is evidence that at least one giant elliptical galaxy contains dark matter at large distances from the optical object.

There is additional evidence that other X-ray galaxies contain dark matter. Observations by William Forman and Christine Forman-Jones of about 50 elliptical galaxies located outside of the cores of clusters reveal that each galaxy has a massive X-ray halo. Using arguments of hydrostatic equilibrium analogous to those used for M87, the astronomers show that each galaxy must contain a significant quantity of dark matter. Although there is still some question of the applicability of the equilibrium arguments, it seems quite probable that elliptical galaxies, like their spiral counterparts, contain more dark than luminous matter.

How much dark matter is there?

How much of the matter in the universe is dark? Let's start by discussing how much matter we can see. When you attempt to build a galaxy out of stars and gas and dust, as Beatrice Tinsley and Richard Larson and others have done, you find that the mean of the ratio of mass-to-luminosity, M/L, is of the order of 1 or 2. In a typical galaxy, for each unit of mass, there is, on the average, one or two units of luminosity. Thus, from their visible spectral properties, most spirals appear to have an average stellar population not too unlike that of the Sun. But the galaxy dynamics tell a different story. Across the optical image of a galaxy, the luminous disk is matched by an equal nonluminous halo mass. Moreover, beyond the optical galaxy out to the largest radii to which rotation velocities have been measured, the dark matter amounts to five or ten times the luminous matter.

To pursue the answer for a larger region of the universe, we must enter the realm of cosmology. In cosmology, observational facts are hard to come by. However, most astronomers do agree on a few that are essential to any theory describing the history and evolution of galaxies. These are:

(1) The universe originated in a Big Bang, whose laws of physics – at least after the first few minutes – are the same laws of physics we know today.

(2) The 3 degrees Kelvin black body radiation is the remnant of the primordial fireball: it has been expanding and cooling ever since. Due to the expansion of the universe, the distance between isolated galaxies is presently increasing. Only within the dense clusters do the mutual gravitational attractions of the galaxies keep these regions from participating in the universal expansion.

(3) Our galaxy (and hence the observations which we make from the Earth) is not at rest with respect to this remnant radiation. But once this motion is accounted for, the background radiation is isotropic on small angular scales to an impressively high degree, a few parts in 100 000. This is generally interpreted to mean that the early universe was smooth and isotropic.

(4) The region over which we observe luminous galaxies is clumpy on enormous scales. Regions of long strings of superclusters separate large regions void of any luminous galaxies. Observers have not yet examined any large volume of space in which the distribution of galaxies is smooth.

In addition, we must define a few terms in order to evaluate the mean density of luminous and dark matter. We accept the following consistent set of values where H_0, the value for the expansion rate of the universe, is uncertain within a factor of 2, but most of the other listed quantities will scale for other values of H_0; G is the gravitational constant; $M_\odot$ is the solar mass; $L_\odot$ is the solar luminosity; and $\langle\varrho\rangle$ is the mean mass density.

$$H_0 = 50 \text{ kilometers per second per cubic megaparsec}$$
$$\varrho = 3H_0^2/8\pi G$$
$$\langle\varrho\rangle = \sim 10^{11}\, M_\odot \text{ per cubic megaparsec}$$
$$\langle L\rangle = \sim 10^8\, L_\odot \text{ per cubic megaparsec}$$
$$\Omega = \varrho/\varrho_c$$
$$M/L = \Omega\varrho_c \simeq \Omega\, 1000\, M_\odot/L_\odot$$

If there is sufficient mass in the universe, the mutual gravitational attraction of all the mass will ultimately halt the expansion; it could even cause the universe to start contracting. The mean mass density which will *just* halt the expansion, i.e., close the universe, is designated ϱ_c (where the c stands for closure or critical). Ω is the ratio of the observed density ϱ to the closure density ϱ_c. The mean mass density of the universe, calculated from gravitational theory, is of order 10^{11} $M_\odot$ Mpc^{-3}, or 1000 times larger than the mean luminosity density, $\langle L\rangle$. The mean luminosity density is a very uncertain number, for it comes from counting up the luminosity from all the faint galaxies, an uncertain guess is best, and it also depends upon the dimming effects of the gas and dust within our galaxy. But as these adopted values show, it will take a value of $M/L = 1000$ to close the universe. Thus, on average, there must be 1000 units of mass for each unit of luminosity. This density corresponds to placing a galaxy 1/10 the mass of the Milky Way in each cubic megaparsec of the universe. But such high densities are not seen optically – nor even from the dynamical studies described above. The dynamical results imply values of M/L of tens and hundreds, but not as high as 1000. Order-of-magnitude results for various dynamical systems are shown in Table 4.1. While exact numbers are controversial, I think this set of values is a fair assessment of the current status of our knowledge.

As is apparent from the upper part of the table, various dynamical analyses give values of M/L less than about 200, corresponding to a value of $\Omega < 0.2$. Thus, all of the dynamical studies would be satisfied with a universe whose mass is only about 1/5 the critical mass, but one in which the quantity of dark matter exceeds by a factor of 10 the luminous matter.

Table 4.1. *Values of M/L and Ω for various dynamical systems*

System	Mass	Scale (kiloparsecs)	M/L	Ω
Visible				
Galaxy	10^9–10^{11}	25	2	0.002
Dynamical				
Galaxy	10^{10}–10^{12}	25	10	0.01–0.02
Binary	10^{10}–10^{13}	50	50	0.05
Group	10^{13}	150	150	0.15
Cluster	10^{14}	250	250	0.25
Local Supercluster	10^{15}	20 000	300	0.15–0.3
Deuterium abundance				<0.2
Inflation				1
Closed				>1

A measure of the density of the universe comes also from the theory of nucleogenesis in the conventional Big Bang cosmology. Because deuterium, produced in the initial Big Bang, can only be destroyed by subsequent evolution, its current abundance is a measure of the density of the universe. Early studies by Tinsley, David Schramm, and coworkers, and recent detailed analysis by Jean Audouze and his colleagues, are all consistent in deducing a low universe density, i.e., $\Omega < 0.2$. Thus, our universe can be one in which the dark matter is 'normal', that is, the type that makes up galaxies, stars, and atoms. This matter, termed baryonic, has been processed in stars and has evolved along with the universe. Such a universe is both consistent with all of the dynamical observations and will expand forever.

However, such an 'open' universe poses questions which theorists find difficult to answer. How did galaxies initially form in such a universe? Why is the microwave background radiation so smooth, given the lumpy distribution of matter in a universe so large that regions of it could never have communicated with each other? And, why is $\Omega = 0.2$ so tantalizingly close to $\Omega = 1$? Why, for example, is Ω observed not either orders of magnitude smaller or larger than unity?

Given these valid questions, theorists have postulated a universe in which $\Omega = 1$. In such a universe, the amount of dark matter is five times larger than that required by the dynamical arguments; and, by

DENNIS the MENACE

"LOTS OF THINGS ARE INVISIBLE, BUT WE DON'T KNOW HOW MANY BECAUSE WE CAN'T **SEE** THEM."

Figure 4.14 Dennis the Menace ® used by permission of Hank Ketcham and © by North America Syndicate.

the limits imposed by nucleogenesis, this dark matter cannot be baryonic. As Alan Guth will explain elsewhere in this volume, these $\Omega = 1$ models modify conventional Big Bang cosmology to incorporate a time of enormously rapid inflation during the initial universe, thus solving the smoothness problem. Such a universe will not expand forever, but will slowly coast to a halt. Only if $\Omega > 1$ will the universe ultimately contract.

We may simplify the current ideas concerning models of the universe into these two extremes: one with $\Omega = 0.2$, which is derived from dynamical arguments, and one with $\Omega = 1$, which is derived from theoretical arguments. There is, however, still a third model, which is presently favored by only a very few. In this model, 'what you see is what you get.' That is, the distribution of mass in a galaxy is described by the distribution of light, but Newtonian potential theory is assumed to be not valid for the very low values of acceleration encountered. In this model, Equation (1) cannot be the correct form for analyzing the distribution of mass in a galaxy; rather, the observed distribution of light gives rise to the observed velocities. Consequently, there is no invisible matter. The mean density of the universe is just that which is visible, and is therefore very small, $\Omega = 0.002$ or so. Most astronomers prefer to accept a universe filled with dark matter rather than to alter Newtonian gravitational theory. Yet Mordehai Milgrom and Jacob Bekenstein, two proponents of a modified theory, point out that laws of physics have been altered in the past on the basis of evidence weaker than the evidence that has led astronomers and physicists to postulate the existence of invisible matter.

We conclude with an uncertainty: there is significantly more dark matter than luminous matter in the universe. But whether that dark matter comprises 90 or 99 percent of mass is unknown; whether the universe will continue to expand forever is unknown. Even Dennis the Menace (Figure 4.14) recognized the problems in determining how much dark matter there is.

What is the dark matter?

What is the dark matter made of? Answers to this question contain many 'ifs'. If we live in a low-density, $\Omega \sim 0.2$ universe, it could all be baryonic, that is, the normal stuff that makes stars and galaxies and us. And, if it is baryonic, it may be faint brown dwarfs, small stars which never became hot enough to start conventional nuclear processing. Or a large population of undetected white dwarfs. Or enormous numbers of cold planet-like objects. Or mini-black holes. Or even maxi-black holes – remnants of the early universe.

If the dark matter is not baryonic, that is, not formed from neutrons and protons, then the large-scale distribution of galaxies in the universe is a clue as to its nature. If galaxies formed in a universe of exotic particles, these particles determined the dynamical evolution of the early universe, and by extension, the nature of the universe we inhabit today. A signature of this early universe must still be present in the large-scale distribution of galaxies which we currently observe.

The plotting of the million brightest galaxies (Figure 4.8), illustrates a complex distribution of chains, strings, filaments, and voids, all connected in a lace-like pattern. This plot is visual evidence of the nonrandom distribution of bright galaxies. But not only is the two-dimensional distribution of galaxies clumped; velocities are also clumped. Radial velocity measures of galaxies observed beyond the constellation of Bootes by Kirshner, Oemler, Schechter, and Schectman offered impressive evidence that there is a region (in velocity space) void of galaxies, which extends over 160 million light years. These observations, coupled with the identification of other large regions void of galaxies, have produced an enormous industry in N-body calculations, as astronomers attempt to identify the physical and dynamical conditions which would give rise to such clumpy distributions. The deepest survey to date of galaxy distribution is that of Huchra, Geller, and de Lapparent described earlier. They obtained velocities for all galaxies brighter than magnitude 15.5 in a strip of sky 6° by 120° going through the Coma Cluster. Assuming that the Hubble flow is smooth, an assumption which has yet to be seriously tested by independent distances to galaxies at large distances, these results show the galaxies arranged around the peripheries of giant voids, forming structures as large as the scale of the observations. These enormous features may be coming close to violating the smoothness observed in the microwave background. They certainly tax standard gravitational models. They may instead imply explosive hydrodynamical formation, as Ostriker and Cowie suggested some years ago. They also allow the possibility that structure size may continue to grow with sample size; we may not yet have observed a large enough fraction of the universe to comprehend its structure.

Currently, we do not even know if anything *is* present in the voids. Are voids empty of all matter? Or only void of baryonic matter? Are low-luminosity galaxies present there? Is there a critical density, below which galaxies will not form? Does the formation of some galaxies inhibit the formation of others? These are some of the questions which follow from the apparently clumpy distribution of the most luminous galaxies.

Nonbaryonic forms of matter that have been suggested include neutrinos, gravitinos, photinos, sneutrinos, axions, magnetic monopoles, and dozens more. The cosmology predicted by each of these particles is complex, and none of them has properties which predict all of the observations. Neutrinos were one of the first appealing candidates for the dark matter. Neutrinos are known to exist; they have generally been assumed to have zero mass. Ramanath Cowsik and John McClelland pointed out in 1973 that if the neutrino had a mass in the range of tens of electron volts, this would have interesting cosmological implications. In the years since then, experimental evidence for the existence of a neutrino mass has been questioned. Most recently, the detection on Earth of neutrino events which arose from the supernova in the Large Magellanic Cloud has reopened the discussion of the neutrino's mass.

But neutrinos are hot, relativistically moving particles, and a universe dominated by neutrinos would form enormous structures early in its history. Such a universe would produce a 'top-down' cosmology, in which the largest structures form first, substructures form clusters next, and galaxies form last. Such a cosmology has interesting large-scale properties, which match well the clusters and superclusters we see as long strings in the plots of galaxy distribution (Figure 4.8). Despite this promise, the initial enthusiasm for a neutrino dominated universe has faded. First, it is still not established that the neutrino has mass. Second, fragmentation of the largest structures down to the sizes of galaxies would take an appreciable fraction of the age of the universe and galaxies would have only recently formed. This timescale seems incompatible with our present ideas of galaxy evolution.

Hence, many cosmologists now favor an alternative model, one in which the particles dominating the universe are cold, rather than hot. Such particles, of which the axion is an oft chosen one, have never been detected, but they are allowed by the physics of the theories from which they emerge. Early in such a universe, the cold axions form clouds which withstand the expansion and clump on all sizes, from clusters of stars to clusters of galaxies. Because fluctuations on all scales condense at the same time, many of the curious interactions between galaxies and their environment can be understood. The overwhelming drawback of such models is that they are based on particles whose existence is currently only postulated. Laboratory experiments presently underway are attempting to detect the axion. Particle physics and astronomy will grow even closer with a detection of such exotic particles.

Those of us curious about matter in the universe hope for better answers to the questions: What is it? And, how much is there? Whatever it is, it must be dark, it must clump about galaxies, it must be less concentrated to the centers of galaxies than is the light, and it must not appreciably obscure the background galaxies. Astronomers are fond of saying that the dark matter could be cold planets, dead stars, bricks, or baseball bats. Physicists are fond of saying that it could be billions of mini-black holes, or somewhat fewer maxi-black holes, or, indeed, any one of a number of exotic particles from the zoo of objects permitted by physical theories but never yet observed. Whatever it is – and it could be of more than one type – it must be the major constituent of our universe.

In a very real sense, astronomy begins anew. The joy and fun of understanding the universe we bequeath to our grandchildren – and to their grandchildren. With over 90 percent of the matter in the universe still to play with, even the sky will not be the limit.

5

Starting the universe: the Big Bang and cosmic inflation

ALAN H. GUTH
*Massachusets Institute of Technology
and
Harvard–Smithsonian Center for Astrophysics*

In the late 1970s, I joined a small drove of particle theorists who began to dabble in studies of the early universe. We were motivated partly by the intrinsic fascination of cosmology, but also by new developments in elementary particle physics itself.

The particle physics motivation arose primarily from the advent of a new class of theories, known as 'grand unified theories' (GUTs†). These theories were invented in 1974, but it was not until about 1978 that they became a topic of widespread interest in the particle physics community. Spectacularly bold, these theories attempt to extend our understanding of particle physics to energies of about 10^{14} GeV (1 GeV = 1 billion electron volts $\approx$ rest energy of a proton). This amount of energy, by the standards of your local power company, may not seem so impressive – it is about what it takes to light a 100-watt bulb for 1 minute. The GUTs, however, attempt to describe what happens when that much energy is deposited on a single elementary particle. Such an extraordinary concentration of energy exceeds the capabilities of the largest existing particle accelerators by eleven orders of magnitude.

To get some feeling for how high this energy really is, imagine trying to build an accelerator that might reach these energies. One can do it in principle by building a very long linear accelerator. The largest existing linear accelerator is at Stanford, with a length of about 2 miles and a maximum energy of about 40 GeV. The output energy is

†I am trying to persuade my colleagues that the word 'theory' is more properly abbreviated as 'TH'. I must admit, however, that this improved orthography has not been widely adopted.

proportional to the length, so a simple calculation shows how long an accelerator would have to be to reach an energy of 10^{14} GeV. The answer is almost exactly 1 light year!

The Department of Energy, unfortunately, seems to be very unreceptive to proposals for funding a 1-light-year-long accelerator. Consequently, if we want to see the most dramatic new implications of the GUTs, we are forced to turn to the only laboratory which has ever reached these energies. That 'laboratory' appears to be the universe itself, in its very infancy. According to the standard hot Big Bang theory of cosmology, the universe had a temperature corresponding to a mean thermal energy of 10^{14} GeV at a time of about 10^{-35} seconds after the Big Bang. No wonder particle theorists suddenly became interested in the very early universe.

The first half of this chapter will review the standard hot Big Bang model of the early universe, while the second half will discuss the developments that have taken place since 1978 – developments which have been motivated mainly by ideas from particle physics.

The Big Bang theory

Cosmology in the twentieth century began with the work of Einstein. In March 1916, Einstein completed a landmark paper titled 'The Foundation of the General Theory of Relativity.' The theory of general relativity is, in fact, nothing more nor less than a new theory of gravity. Complex but very elegant, the theory describes gravity as a distortion of the geometry of space and time. Unlike Newton's theory of gravity, general relativity is consistent with the ideas of 'special relativity,' which Einstein had introduced in 1905. While the rest of the world waited to be persuaded, Einstein was immediately convinced that he had found the correct description of gravity.

In less than a year after the publication of general relativity, Einstein applied it to the universe as a whole. However, in carrying out these studies, Einstein discovered something that surprised him a great deal: It was impossible to build a static model of the universe consistent with general relativity. Einstein was perplexed by this fact. Like his predecessors, he had looked into the sky, saw that the stars appeared motionless, and erroneously concluded that the universe is static.

In fact, the same problem that Einstein discovered in the context of general relativity also existed in Newtonian mechanics, although it had not been appreciated until the work of Einstein. The problem is fairly simple to understand: If masses were distributed uniformly and statically throughout space, then everything would attract everything else and the entire configuration would collapse.

Einstein nonetheless remained convinced that the universe was static. He therefore modified his equations of general relativity, adding what he called a 'cosmological term' – a kind of universal repulsion that prevents the uniform distribution of matter from collapsing under the normal force of gravity. The cosmological term, Einstein found, fits neatly into the equations of general relativity – it is completely consistent with all the fundamental ideas on which the theory was constructed.

Einstein's ideas remained viable for about a decade, until astronomers began to measure the velocities of distant galaxies. They then discovered that the universe is not at all static. To the contrary, the distant galaxies are receding from us at high velocities.

The pattern of the cosmic motion was codified at the end of the 1920s by Edwin Hubble, in what we now know as Hubble's Law. Hubble's Law states that each distant galaxy is receding from us with a velocity which is, to a high degree of accuracy, proportional to its distance. Thus one can write

$$v = Hd$$

where v is the recession velocity, d is the distance to the galaxy, and the quantity H is known as the Hubble 'Constant.' (I put the word 'Constant' in quotation marks to call attention to its inaccuracy. The quantity was called a 'constant' by astronomers, presumably because it remains approximately constant over the lifetime of an astronomer. The value of H changes, however, as the universe evolves; so, from the point of view of a cosmologist, it is not a constant at all.)

The value of the Hubble Constant is not very well known. The recession velocities of the distant galaxies are no problem – they can be determined very accurately from the Doppler shift of the spectral lines in the light coming from the galaxies. The distances to the galaxies, on the other hand, are very difficult to determine, as both Robert Kirshner and James Gunn note. These distances are estimated by a variety of indirect methods, and the resulting value of the Hubble Constant is thought to be uncertain by a factor of about 2. It is believed to lie somewhere in the range

$$H \approx \frac{0.5 \text{ to } 1}{10^{10} \text{ years}}$$

Notice that the Hubble Constant has the units of inverse time; when the Hubble Constant is multiplied by a distance, the result has the units of distance per time, or velocity. In particular, if the expression for H above is multiplied by a distance in light years, the result is a velocity measured as a fraction of the velocity of light. Alternatively, one can use

$H \approx$ (15 to 30) kilometers per second per million light years

to obtain an answer in kilometers per second.

The development of cosmology in the twentieth century was somewhat confused by the fact that Hubble badly overestimated the value of the Hubble Constant, reporting a value of 150 kilometers per second per million light years. This mismeasurement had important consequences. In the context of the Big Bang model, an erroneously high value for the expansion rate implies an erroneously low value for the age of the universe. Hubble's value for the Hubble Constant implied an age of about 2 billion years, a number which conflicts with geological evidence that the Earth is significantly older. Not until 1958 did the measured value of the Hubble Constant come within the currently accepted range, due primarily to the work of Walter Baade and Allan Sandage. The age of the universe is now estimated to be between 10 and 20 billion years.

Once it is noticed that the other galaxies are receding from us, there are two conceivable explanations. The first is that we might be in the center of the universe, with everything moving radially outward from us like the spokes on a wheel. In the early sixteenth century such an explanation would have been considered perfectly acceptable. Since the time of Copernicus, however, astronomers and physicists have become instinctively skeptical of this kind of reasoning, so alternative explanations are sought. In this case, an attractive alternative can be found.

The alternative explanation can be called *homogeneous expansion*,

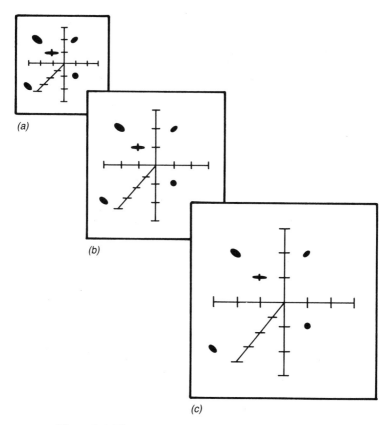

Figure 5.1 The expanding homogeneous universe. The three pictures show snapshots of the same region of the universe, taken at three successive times. Each picture is essentially a photographic blowup of the previous picture, with all distances enlarged by the same percentage. An observer on any galaxy would conclude that all the other galaxies are receding from him.

and it is illustrated in Figure 5.1. The three pictures are intended to show successive snapshots of a region of the universe. Each picture is essentially a photographic blowup of the previous picture, with all distances enlarged by the same percentage. According to this explanation, there is nothing special about our galaxy – or any other galaxy. All galaxies are approximately equivalent, and are spread more or less uniformly throughout all of space. (In this description, there is no center and no edge to the distribution of galaxies.) As the system evolves, from diagram (*a*) to (*b*) to (*c*), *all* intergalactic distances are enlarged. Thus, regardless of which galaxy we are living on, we would see all the other galaxies receding from us. Furthermore, this picture leads immediately to the conclusion that the recession velocities obey Hubble's Law. Since all distances increase by the same *percentage* as the system evolves, larger distances naturally increase by a larger amount. The apparent velocity of a galaxy is proportional to the amount by which the distance from us increases, and hence it is proportional to the distance.

When one extrapolates this picture backwards in time, one finds an instant in the past when all the galaxies must have been on top of each other and when the density of the universe would have been infinite. Such an event is called a 'singularity;' and, in this case, the singularity is the instant of the 'Big Bang' itself, some 10–20 billion years ago. The time of the Big Bang is very uncertain for two reasons: We do not know Hubble's Constant very accurately, and we are also uncertain about the mass density of the universe. (The mass density is important in calculating the history of the universe, because it determines how fast the cosmic expansion is slowing down under the influence of gravity.) Both these problems have been addressed in detail by other authors in this book.

The reader also should be warned that the calculation implying an infinite density at the instant of the Big Bang is not to be trusted. As one looks backward in time with the density going up and up, one is led further from the conditions under which the laws of physics as we know them were developed. Thus, it is quite likely that at some point these laws become totally invalid, and then it is a matter of guesswork to discuss what happened at earlier times. Nonetheless, as the history of the universe is extrapolated backward, the density increases without limit for as long as the known laws of physics apply. Indeed, most cosmologists today are reasonably confident in our understanding of the history of the universe back to one microsecond (10^{-6} second) after the Big Bang. The goal of the cosmological research involving grand unified theories is to solidify our understanding back to 10^{-35} seconds after the Big Bang.

Having discussed the key features of the Big Bang theory, we can now ask what evidence can be found to support it. In addition to

Hubble's Law, there are two significant pieces of observational evidence in favor of the Big Bang theory.

The first is the observation of the cosmic background radiation. To understand the origin of this radiation, begin by recalling that the temperature of a gas rises when the gas is compressed. For example, a bicycle tire is warmed when it is inflated by a hand pump. Similarly, a gas cools when it is allowed to expand. In the Big Bang model, the universe has been expanding throughout its history, and that means that the early universe must have been much hotter. (In fact, a purely mathematical calculation would suggest that the temperature was infinite at the instant of the Big Bang. This calculation, however, like the calculation of the infinite density, should not be considered convincing.)

All hot matter emits a glow, just like the glow of hot coals in a fire. Thus, the early universe would have been permeated by the glow of light emitted by the hot matter. As the universe expanded, this light would have redshifted. Today, the universe would still be bathed by the radiation, a remnant of the intense heat of the Big Bang – now redshifted into the microwave part of the spectrum. This prediction was confirmed in 1964 when Arno A. Penzias and Robert W. Wilson of the Bell Telephone Laboratories discovered a background of microwave radiation with an effective temperature of about 3 degrees Kelvin.

At the time of their discovery, Penzias and Wilson were not looking for cosmic background radiation. Instead, they were searching for astronomical sources capable of producing low-level radio interference. They discovered a hiss in their detector which they carefully tracked down, verifying that it was an external source of radiation and not simply electrical noise from the receiver. They also found that the radiation arrived at the Earth uniformly from all directions in the sky. Later the spectrum of the radiation was measured, and it was found to agree exactly with the kind of thermal radiation that would be expected from the glow of hot matter in the early universe. A graph of the spectral content of the cosmic background radiation is shown in Figure 5.2.

The second important piece of evidence supporting the Big Bang theory is related to calculations of what is called 'Big Bang nucleo-

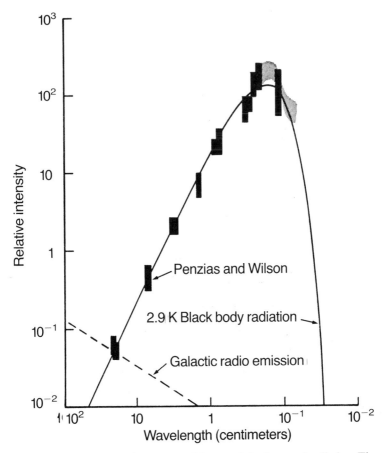

Figure 5.2 The spectrum of the cosmic background radiation. The vertical bars and the shaded region show the results of intensity measurements at a variety of wavelengths. The solid line is a theoretical curve, corresponding to thermal radiation at a temperature of 2.9 degrees Kelvin. (Figure adapted from *The Big Bang* by Joseph Silk. © 1980 W. H. Freeman and Company. Reprinted with permission.)

synthesis.' This evidence is somewhat more difficult to understand than the cosmic background radiation, and it is therefore much less discussed in popular scientific literature. To make any sense out of this argument, one must understand that the Big Bang theory is not just a cartoon description of how the universe may have behaved. On the contrary, the Big Bang theory is a very detailed model. Once one accepts its basic assumptions, then knowledge of the laws of physics allows one to calculate how fast the universe would have expanded, how fast the expansion would have been slowed by gravity, how fast the universe would have cooled, and so on. Given this information, knowledge from nuclear physics allows one to calculate the rates of the different nuclear reactions that took place in the early history of the universe.

The early universe was very hot, so hot that even nuclei would not have been stable. At 2 minutes after the Big Bang there were virtually no nuclei at all. The universe was filled with a hot gas of photons and neutrinos, with a much smaller density of protons, neutrons, and electrons. (The protons, neutrons, and electrons were very unimportant at the time, but they later became raw materials for the formation of stars and planets.) As the universe cooled, the protons and neutrons began to coalesce to form nuclei. From the nuclear reaction rates, one can calculate the expected abundances of the different types of nuclei that would have formed. One finds that most of the matter in the universe would remain in the form of hydrogen. About 25 percent (by mass) of the matter would have been converted to helium, and trace amounts of other nuclei would also have been produced.

Most of the types of nuclei that we observe in the universe today were produced much later in the history of the universe, in the interiors of stars and in supernova explosions. The lightest nuclei, however, were produced primarily in the Big Bang, and it is possible to compare the calculated abundances with direct observations. Such a comparison can be carried out for the abundances of helium-4, helium-3, hydrogen-2 (otherwise known as deuterium), and lithium-7.

The comparison is complicated by the fact that we do not know all the information necessary to carry out the calculations. In particular, the calculation depends on the density of protons and neutrons in the universe, a quantity which can be estimated only roughly by astronomical observations. Thus, the calculations have to be carried out for a broad range of values for this density. One then asks whether there exists a plausible value for which the answers turn out right.

The results of this comparison are shown in Figure 5.3. The horizontal axis shows the present density of protons and neutrons; the curves indicate the results of the calculation. The observations, with the estimated range of their uncertainties, are shown by the shaded

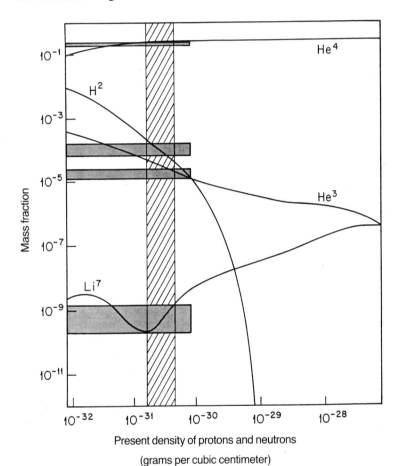

Present density of protons and neutrons

(grams per cubic centimeter)

Figure 5.3 Big Bang nucleosynthesis and the abundances of the light chemical elements. The curves show how the predicted abundances of helium-4, helium-3, deuterium (hydrogen-2), and lithium-7 depend on the present density of protons and neutrons. The shaded horizontal bars show the observed abundances, with the estimated uncertainties. Excellent agreement is obtained for the range of densities indicated by the cross-hatched region. (Figure adapted from D. N. Schramm, *Physics Today*, April 1983.)

horizontal bars. Note that there is a range of values for the density of protons and neutrons, indicated by the cross-hatched region, for which each of the four calculated curves agrees with the corresponding observations. Even though the abundances of these nuclei are not known to high precision, the success of the comparison is very impressive. Note that the abundances span nine orders of magnitude. If there was never a Big Bang, there would be no reason whatsoever to expect that helium-4 would be 10^8 times as abundant as lithium-7 – it might just as well have been the other way around. But, when calculated in the context of the Big Bang theory, the ratio works out just right.

Unanswered questions

Each piece of evidence discussed above – Hubble's Law, the cosmic background radiation, and the Big Bang nucleosynthesis calculations – probes the history of the universe at a different period of time. The observation of Hubble's Law, for example, probes the behavior of the universe at times comparable to the present – billions of years after the Big Bang. The cosmic background radiation, on the other hand, samples the conditions in the universe about 100 000 years after the Big Bang, when the universe became cool enough for the plasma of free nuclei and electrons to condense into neutral atoms. The plasma that filled the universe at earlier times was almost completely opaque to photons, which would have been constantly absorbed and reemitted. With the formation of neutral atoms, however, the universe became highly transparent. Thus, most of the photons in the cosmic background radiation have been moving in a straight line since 100 000 years after the Big Bang, and they therefore provide an image of the universe at that time. Finally, the Big Bang nucleosynthesis calculations probe the history of the universe at much earlier times. The processes involved in establishing the abundances of the light nuclei occurred at times ranging from about 1 second to about 4 minutes after the Big Bang.

The Big Bang theory is a very successful description of the evolution of the universe for the whole range of times discussed above – from about 1 second after the Big Bang to the present. Nonetheless, the standard Big Bang theory has serious shortcomings, in that a

Table 5.1. *Questions left unanswered by the standard Big Bang theory*

#1:	Is there some way of understanding why the ratio of the number of photons to the number of protons and neutrons is about equal to 10^{10}, rather than some other number?
#2:	How did the universe become so homogeneous on large scales? Do we have to assume that it started out that way?
#3:	Why was the mass density of the early universe so extraordinarily close to the critical density?
#4:	Can one find a physical origin for the primordial density perturbations which lead to the evolution of galaxies and clusters of galaxies? Are there physical processes which determine the spectrum of these perturbations?

number of very obvious questions are left unanswered. Here I describe four of these questions, which are listed in Table 5.1. Later I will show how ideas from particle physics have led to a radically new picture for the very early behavior of the universe, a picture that provides plausible answers to each of these questions.

The first question involves the number of protons and neutrons in the universe, relative to the number of photons. Photons are found mainly in the cosmic background radiation, while protons and neutrons form the atomic nuclei of the matter that makes up the galaxies. The observed universe contains about 10^{10} photons for every proton or neutron. The standard Big Bang theory does not explain this ratio, but instead assumes that the ratio is given as a property of the initial conditions.

The second question is related to the large-scale homogeneity, or uniformity, of the observed universe. The discussion of homogeneity must be qualified, however, because the universe that we observe is in many ways very inhomogeneous. The stars, galaxies, and clusters of galaxies make a very lumpy distribution, and the observed universe contains many pronounced inhomogeneities. Cosmologically speaking, however, all of this structure in the universe is very small-scale, even the great 'bubbles and voids' described by Margaret Geller. If one averages over very large scales, scales of 300 million light years or more, then the universe appears to be very homogeneous. This large-scale homogeneity is most evident in the cosmic background radiation. Physicists have probed the temperature of the cosmic background radiation in different directions, and have found it to be extremely uniform. It is just slightly hotter in one direction than in the opposite direction, by about 1 part in 1000. Even this small discrepancy, however, can be accounted for by assuming that the solar system is moving through the cosmic background radiation at a speed of about 600 kilometers per second. Once the effect of this motion is subtracted out, then the resulting temperature pattern is uniform in all directions to the best accuracy that has so far been attained – an accuracy of a few parts in 100 000. Since the cosmic background radiation gives an image of the universe at 100 000 years after the Big Bang, one concludes that the universe was very homogeneous at that time. (The observed 'small-scale' inhomogeneities are believed to have formed later, by the process of gravitational clumping.) The standard Big Bang theory cannot explain the large-scale uniformity; instead, the uniformity must be postulated as part of the initial conditions.

The difficulty in explaining the large-scale uniformity is a quantitative question, related to the rate of expansion of the universe. Under many circumstances a uniform temperature would be easy to understand – anything will come to a uniform temperature if it is left undisturbed for a long enough period of time. In the standard Big Bang theory, however, the universe evolves so quickly that it is impossible for the uniformity to be created by any physical process. In fact, the impossibility of establishing a uniform temperature depends on none of the details of thermal transport physics, but instead is a direct consequence of the principle that no information can propagate faster than the speed of light. One can pretend, if one likes, that the universe is populated with little purple creatures, each equipped with a furnace and a refrigerator, and each dedicated to the cause of trying to create a uniform temperature. However, those busy little creatures would have to communicate at more than *90 times the speed of light* if they are to achieve their goal of creating a uniform temperature across the visible universe by 100 000 years after the Big Bang.

The puzzle of explaining why the universe appears to be uniform over such large distances is not a genuine inconsistency of the standard theory: If the uniformity is assumed in the initial conditions, then the universe will evolve uniformly. The problem is that one of the most salient features of the observed universe – its large-scale uniformity – cannot be explained by the standard Big Bang theory; it must be assumed as an initial condition.

The third question is related to the mass density of the universe. This mass density is usually measured relative to a benchmark called the 'critical mass density,' which is defined in terms of the expansion of the universe. If the mass density exceeds the critical density, then the gravitational pull of everything on everything else will be strong enough to halt the expansion eventually. The universe would recollapse, resulting in what is sometimes called a 'big crunch.' On the other hand, if the mass density is less than the critical density, the universe will go on expanding forever.

Cosmologists typically describe the mass density of the universe by a ratio designated by the Greek letter Ω, defined by

$$\Omega \equiv \frac{\text{Mass density}}{\text{Critical mass density}}$$

Ω is very difficult to determine, but its present value is known to lie somewhere in the range of 0.1–2.

That seems like a broad range, but consideration of the time development of the universe leads to a different point of view. $\Omega = 1$ is an *unstable equilibrium point* of the evolution of the standard Big Bang theory, which means it resembles the situation of a pencil balancing on its sharpened end. The phrase 'equilibrium point' implies that if Ω is ever exactly equal to 1, it will remain exactly equal to 1 forever – just as a pencil balanced precisely on end will, according to the laws of the classical physics, remain forever in the vertical position. The word 'unstable' means that any deviation from the equilibrium point, in either direction, will rapidly grow. If the value of Ω in the early universe is just a little bit above 1, it will rapidly rise toward infinity; if Ω in the early universe is just a tiny bit below 1, it will rapidly fall toward zero. Thus, it seems very unlikely that the value of Ω today would lie *anywhere* in the vicinity of 1.

In fact, for Ω to be anywhere near 1 today, it must have been extraordinarily close to 1 at early times. For example, we can consider the time of 1 second after the Big Bang, the time at which the processes related to Big Bang nucleosynthesis were beginning to take place. In order for Ω to be somewhere in the allowed range today, at 1 second after the Big Bang Ω had to have been equal to 1 to an accuracy of 15 decimal places. If we go further and consider the time of 10^{-35} seconds after the Big Bang, when thermal energies were typical of the energy scale of GUTs, then at that time Ω had to have been equal to 1 to an accuracy of 49 decimal places!

In the standard Big Bang theory there is no explanation whatever for this fact, as has been emphasized by Robert H. Dicke and P. James E. Peebles of Princeton University. At 1 second after the Big Bang, Ω could have had any value – except that most possibilities would lead to a universe very different from the one in which we live. Like the large-scale homogeneity, the nearness of the mass density to the critical density cannot be explained; instead, it must be postulated as part of the initial conditions.

The fourth question concerns the origin of the density perturbations that are responsible for the development of the small-scale inhomogeneities. While the universe is remarkably homogeneous on the very large scales, there is nonetheless a very complicated structure on smaller scales, ranging from solar systems to clusters of galaxies. The existence of this structure is undoubtedly related to the gravitational instability of the universe: Any region which contains a higher-than-average mass density will produce a stronger-than-average gravitational field, thereby pulling in even more excess mass. Thus, small perturbations are amplified to become large perturbations.

However, in order for galaxies to evolve, the early universe must have contained primordial density perturbations. The standard Big Bang model offers no explanation for either the origin or the form of these perturbations. Instead, an entire spectrum of primordial perturbations must be assumed as part of the initial conditions.

Grand unified theories

The four questions listed in Table 5.1 involve some of the most basic and obvious features of the universe, yet the standard Big Bang theory leaves all of these questions unanswered. During the last decade, however, cosmologists have made use of new ideas from elementary particle theory to develop new theories about the behavior of the universe at very early times. In the process, they have discovered plausible answers to each of these questions. Before discussing these new ideas in cosmology, however, it is necessary first to summarize the recent advances in elementary particle physics.

Physicists use the word 'interaction' to refer to any process that elementary particles can undergo, whether it involves scattering, decay, particle annihilation, or particle creation. All of the known interactions of nature are divided into four types. From the weakest to the strongest, these interactions are: gravitation, the weak interactions, electromagnetism, and the strong interactions. *Gravitation* appears to be strong in our everyday lives because it is long-range and universally attractive – thus we are accustomed to feeling the force that acts between all the particles in the Earth and all the particles in our own bodies. The force of gravity acting between two elementary particles, however, is incredibly weak. It is much weaker than any of the other known forces – in fact, so weak it has never been detected. Although the *weak interactions* are much stronger than gravity, they are not noticed in our everyday lives because they have a range that is roughly 100 times smaller than the size of an atomic nucleus. They are seen primarily in the radioactive decay of many kinds of nuclei, and they are also responsible for the scattering of particles called neutrinos, a type of experiment that is now routinely carried out at high-energy accelerator laboratories. *Electromagnetism* includes both electric and magnetic forces, and is responsible for holding the electrons of an atom to the nucleus. Light waves, radio waves, microwaves, and X-rays are also electromagnetic phenomena. Finally, the *strong interactions* have a range of about the size of an atomic nucleus, and they account for the force that binds the protons and neutrons inside a nucleus. They also account for the tremendous energy release of a hydrogen bomb, as well as the interactions of many short-lived particles that are investigated in particle accelerator experiments.

The strong, the weak, and the electromagnetic interactions all appear to be accurately described by theories developed during the early 1970s. The strong interactions are described by *quantum chromodynamics*, or QCD, a theory based on the hypothesis that all strongly interacting particles are composed of quarks. The theory provides a detailed description of the interactions that bind the quarks into the observed particles, and the residual effect of these quark interactions can account for the observed interactions of the particles. Unfortunately, our ability to extract quantitative predictions from QCD is very limited. The theory is very intricate, and at present only some of its consequences can be reliably calculated. Nonetheless, the evidence for QCD is strong enough that most particle physicists are convinced that the theory correctly describes the strong interactions over the full range of available energies.

The weak and electromagnetic interactions are successfully described by the unified electroweak theory, also known as the Glashow–Weinberg–Salam theory (named for Sheldon Glashow of Harvard University, Steven Weinberg of the University of Texas at Austin, and Abdus Salam of the International Center for Theoretical Physics in Trieste, who shared the 1979 Nobel Prize in Physics for this work). Standard calculational techniques are very effective in extracting predictions from this theory, owing to the inherent weakness of the interactions being described.

While a quantum theory of gravity remains to be developed, we nonetheless believe that general relativity provides the correct description of gravity at the level of classical physics – that is, in the approximation that the effects of quantum theory can be ignored. The effects of gravity, however, are noticeable only when the number of elementary particles is very large. The classical approximation is incredibly accurate in these situations, and the theory of general relativity is therefore sufficient to describe all the observed properties of gravity.

QCD and the unified electroweak theory, when taken together, have come to be called the *standard model of elementary particle physics*. Embedded in the standard model are three different types of fundamental interactions, labeled by the symbols $U(1)$, $SU(2)$, and $SU(3)$. (These symbols are actually the names of mathematical symmetry groups which determine the form of the interactions, but for our purposes the symbols can be taken simply as labels for the three interactions.) The $U(1)$ and $SU(2)$ interactions are the fundamental ingredients of the Glashow–Weinberg–Salam theory, and they combine together in a somewhat complicated way to describe the weak and electromagnetic interactions. The $SU(3)$ label refers to the strong interactions described by QCD.

Thus, since the early 1970s, elementary particle physics has been in a state of unprecedented success. The electromagnetic, weak, and strong interactions are successfully described by the standard model of particle physics in terms of three fundamental interactions. In other words, all known physics can be described by the standard model of particle physics and/or the theory of general relativity.

GUTs emerged from this atmosphere of enormous success. The first grand unified theory, called the 'minimal $SU(5)$ model,' was proposed in 1974 by Glashow and Howard Georgi, also of Harvard University.

The basic idea of grand unification is that the $U(1)$, $SU(2)$, and $SU(3)$ interactions are actually components of a single unified force. At first this idea seems impossible, since the strengths of the three types of interactions are very different. The interaction strengths cannot be determined theoretically, but instead must be fixed by experiment. The theory, however, implies that the interaction strengths depend on the energy of the particles that are interacting. Once the strengths of the three interactions are measured at one energy, the theory allows one to calculate the strengths at any other energy. The results of such a calculation are shown in Figure 5.4. The important feature is that all three lines appear to meet rather accurately at an energy level somewhere between 10^{14} and 10^{15} GeV. It is this calculation, first carried out by Georgi, Weinberg (then at Harvard University), and Helen R. Quinn (then at Harvard University, now at the Stanford Linear Accelerator Center), that determines the enormous energy scale of the GUTs.

According to GUTs, there is really only one interaction, not three. If we were able to do experiments in the energy range of 10^{14} or 10^{15} GeV, then we would see this clearly. At lower energies, however, there is a mechanism that causes the single interaction to look as if it were three separate interactions. The mechanism is called *spontaneous symmetry breaking*, and I will describe some of its properties in the next section. For now, the reader should be aware only that spontaneous symmetry breaking is not new to GUTs. The mechanism has been used very successfully in the Glashow–Weinberg–Salam theory of the electroweak interactions, and similar phenomena are known to occur in condensed matter physics.

While the standard model of particle physics discussed earlier is very well established, the same cannot be said for GUTs. Even if the idea of grand unification is correct, we certainly do not know which of the many conceivable theories is likely to be the right one. Nonethe-

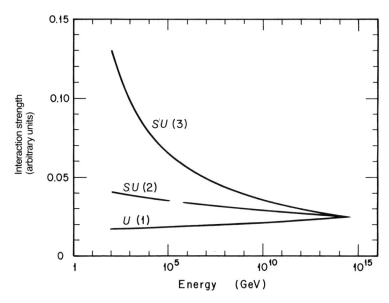

Figure 5.4 Dependence of interaction strengths on energy. The strengths of the three types of fundamental interactions – $U(1)$, $SU(2)$, and $SU(3)$ – are measured at energies of about 100 GeV. The curves show the calculated strengths of these interactions at higher energies, according to the standard model of particle physics. The meeting of all three curves at an energy between 10^{14} and 10^{15} GeV suggests that the three interactions are unified at this energy.

less, GUTs are considered highly attractive for a variety of reasons, and here I would like to explain two of them.

First, GUTs are the only known theories which predict that the charges of the electron and the proton should be equal in magnitude. When I tell this fact to my friends who are not particle physicists, I find that many of them are unimpressed. They may have learned in high school that those two charges are equal, and it never seemed very consequential. High school teachers are notorious, however, for neglecting to tell their students that prior to GUTs, nobody had even a fuzzy idea about *why* those two charges are equal. In all theories developed prior to the GUTs, the two charges could each have been any number whatever; it was just an experimental coincidence that for some reason they happened to be equal to each other to at least 1 part in 10^{20}. GUTs, on the other hand, contain a fundamental symmetry that relates the behavior of electrons to the behavior of the quarks which make up the proton. This symmetry guarantees that the charges are equal. Furthermore, if this symmetry were violated by even the smallest amount, then the theory would no longer be mathematically well defined. Thus, if the charge of the electron were found to differ in magnitude from the charge of a proton by, for example, even 1 part in 10^{24}, then GUTs would have to be abandoned. But, if successively more accurate experiments continue to confirm that the two charges are equal, then such a result would have to be considered as further evidence in favor of grand unification.

A second reason why GUTs are considered attractive involves a topic that I have already discussed, but this time I will describe it more quantitatively. Based on the idea that the three interactions of the standard model of particle physics arise from a single fundamental interaction, GUTs imply that the three curves describing the interaction strengths in Figure 5.4 must all meet at a point. That means, for example, if any two interaction strengths are measured, then the third interaction strength can be predicted by the criterion that its curve must pass through the point where the first two curves crossed each other. This prediction of the GUTs works very well – the experimental result agrees with the prediction to an accuracy of about 3 percent, while the estimated experimental uncertainty of the test is about 6 percent.

To present the full picture, I should mention that GUTs suffer from one important drawback, known as the 'hierarchy' problem. This problem is largely aesthetic, but it is taken quite seriously by the particle physics community. The problem is that the enormous energy scale of GUTs – 10^{14} GeV – has to be 'put in by hand.' When we say the number is 'put in by hand,' we mean that there is no known *a priori* reason why this energy scale is so many orders of magnitude larger than other energy scales in particle physics. Recall, however, that there is a clear *experimental* reason for believing that the energy scale of unification is very high: According to Figure 5.4, the scale of unification must be very high in order to account for the large differences in the strengths of the three interactions observed at the energies of experimental particle physics. To understand the attitude of the particle theorists, one must realize that GUTs are not seen as the ultimate fundamental theory of nature. First of all, the ultimate fundamental theory must obviously include gravity, which GUTs do not. Second, GUTs are considered too inelegant to be serious candidates for the ultimate fundamental theory of nature. In particular, even the simplest of the GUTs contains over 20 free parameters (that is, numbers, such as the charge of an electron, that must be measured experimentally before the theory can be used to make predictions). Thus, the particle theorist expects that someday the correct GUT will be derived as an approximation to the ultimate theory, which will contain few if any free parameters. In this context, the energy scale of grand unification will be calculable. When the particle theorist says that the energy scale of GUTs is 'put in by hand,' he is really saying that he does not at the present time see any reason why this hypothetical calculation of the future will give such a large number. Advocates of GUTs hope someday that reason will be found.

Spontaneous symmetry breaking

Before returning to the discussion of cosmology, I would first like to describe some of the properties of spontaneous symmetry breaking. According to the general definition, a spontaneously broken symmetry is one which is present in the underlying theory describing a system, but which is hidden when the system is in its equilibrium state. To give the reader some familiarity with this concept, I will describe the spontaneous symmetry breaking in a GUT by comparing it with the spontaneous symmetry breaking which occurs in a much more familiar system – a crystal. (This section will nonetheless be somewhat more complicated than the rest of the chapter, so the reader who finds it tedious is invited to skip to the next section. The discussion of spontaneous symmetry breaking will not be necessary to understand what will follow, although the reader who is willing to wade through this section will certainly finish up with a more complete understanding.)

The analogy between spontaneous symmetry breaking in crystals and spontaneous symmetry breaking in GUTs is outlined in Table 5.2. In order to make the analogy as clear as possible, I will discuss a particularly simple type of crystal, a type called 'orthorhombic.' The structure of an orthorhombic crystal is illustrated in Figure 5.5. These crystals have a rectangular structure, so all the angles are right angles.

Table 5.2. *Spontaneous symmetry breaking:*
the crystal–grand unified theory analogy

	Crystal	GUTs
Symmetry	Rotational invariance	Electron, neutrino, and quark indistinguishable Three interactions indistinguishable
Spontaneous symmetry breaking	Crystal axes pick out three distinct directions	Higgs fields pick out three distinct particles – electron, neutrino, and quark – and also three distinct interactions
Low-energy physics	Three fundamental axes of space Three fundamental speeds of light	Three distinct particles Three distinct interactions
High-temperature physics	Crystal melts – rotational invariance restored	Phase transition at $T \approx 10^{27}$ degrees Kelvin – symmetry restored

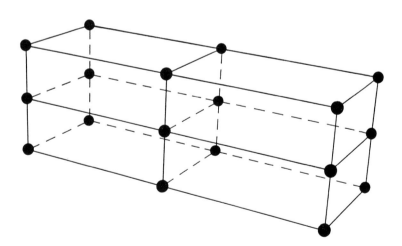

Figure 5.5 The structure of an orthorhombic crystal. This type of crystal has a rectangular structure, in which all three principal lengths are different. The formation of such a crystal is an example of spontaneous symmetry breaking, analogous to the spontaneous symmetry breaking in GUTs.

However, unlike a simple cubic crystal in which all the principal lengths are equal, the three principal lengths of orthorhombic crystals are all different. (This feature of orthorhombic crystals will make the analogy with GUTs a little closer.) A crystal of topaz provides an example of the orthorhombic structure.

Starting at the top of Table 5.2, the first row indicates the symmetry that is involved. For the case of the crystal, the relevant symmetry is rotational invariance – that is, the physical laws describing the system make no distinction between one direction of space and another. For GUTs, the symmetry is more abstract, having nothing to do with orientation in physical space. Instead, the symmetry relates the behavior of one type of particle to the behavior of another; it is therefore sometimes called an 'internal' symmetry. In this case, the underlying symmetry of GUTs implies that the three interactions of the standard model of particle physics – $U(1)$, $SU(2)$, and $SU(3)$ – are really one interaction, and hence indistinguishable. In addition, the symmetry implies that individual particles, which are normally distinguished from each other by how they participate in these interactions, will necessarily lose their identity. In particular, GUTs imply that the underlying laws of physics make no distinction between an electron, a neutrino, or a quark. The indistinguishability of these three particles, as well as the indistinguishability of the $U(1)$, $SU(2)$, and $SU(3)$ interactions, is analogous to the indistinguishability of the different directions of space in the case of the crystal.

The next row of Table 5.2 indicates the nature of the spontaneous symmetry breaking. In the case of the crystal, the atoms arrange themselves along specific crystallographic axes which are picked out randomly as the crystal is formed, and thus three directions of space become distinct. In GUTs, a set of fields is added for the specific purpose of spontaneously breaking the symmetry. These fields are known as Higgs fields (after Peter W. Higgs of the University of Edinburgh), and the spontaneous-symmetry-breaking mechanism, which occurs in a variety of particle physics theories, is known as the Higgs mechanism.

In terms of the structure of the theory, the Higgs fields are on an equal footing with the other fundamental fields, such as the electromagnetic field. It is postulated that these fields exist, and that they evolve according to a specified set of equations. While the electromagnetic field gives rise to photons, the Higgs fields give rise to Higgs particles. The Higgs particles associated with the breaking of the grand unified symmetry are expected to have masses corresponding to energies in the vicinity of 10^{14} GeV, which means that they are far too massive to be produced in the foreseeable future. There is another Higgs particle, however, associated with the spontaneous symmetry breaking in the Glashow–Weinberg–Salam theory, which could conceivably be detected in upcoming accelerator experiments.

The spontaneous symmetry breaking is accomplished by formulating the theory in such a way that the Higgs fields have nonzero values in the vacuum. (To the particle theorist, the word 'vacuum' is defined as the state of lowest possible energy density.) Just as the atoms in the crystal can align equally well along any of an infinite number of possible orientations, the Higgs fields can equally well assume any of an infinite number of combinations of values. Some particular combination of Higgs field values is chosen randomly as the system is formed, just as distinct directions in space are chosen randomly as a crystal begins to form. This random choice of nonzero Higgs field values breaks the grand unified symmetry. The other particles in the theory interact with the Higgs fields, producing the apparent distinction between the $U(1)$, $SU(2)$, and $SU(3)$ interactions, and also the apparent distinction between electrons, neutrinos, and quarks.

The next row of Table 5.2 describes the behavior of low-energy physics in the two systems. Here the analogy can be made more illustrative by whimsically imagining a world of intelligent creatures living inside an orthorhombic crystal. Let us assume that these creatures can somehow move about and carry on the task of scientific investigation, but that they cannot muster enough energy to melt or even significantly perturb the crystal in which they live. In their world, the crystal is not considered an object, but a fundamental property of space. A physics book would make no mention of rotational symmetry, but would instead contain a chapter discussing the properties of space and its primary axes. The crystalline structure would, for example, affect the propagation of light through the medium, and a table of physical constants in the crystal world would list three speeds of light, one for each primary axis. If grand unification is correct, then the world of our experience is similar to this crystal world; our tabulation of the different properties of the strong, weak, and electromagnetic interactions is analogous to the tabulation of the different speeds of light. Similarly, the distinct properties that we observe for electrons, neutrinos, and quarks are not fundamental – they represent the different ways that particles can interact with the fixed 'Higgs field crystal' in which we live.

Finally, the last row of Table 5.2 describes the high-temperature behavior of the two systems. If a crystal is heated sufficiently, it will undergo a phase transition (that is, it will melt) to become a liquid. The distribution of molecules in the liquid is rotationally symmetric, looking the same no matter how the liquid is turned. Thus at high temperatures, the rotational symmetry is restored. According to GUTs, an analogous phase transition occurs at a temperature of the order of 10^{27} degrees Kelvin. (This temperature corresponds to a mean thermal energy of – you guessed it – 10^{14} GeV.) At temperatures higher than this value, the 'Higgs field crystal' in which we live would enter a different phase. The Higgs fields would oscillate wildly under the thermal agitation, but the mean value of each field would be zero, so the grand unified symmetry would be restored. In this phase, the $U(1)$, $SU(2)$, and $SU(3)$ interactions would all merge into a single interaction, and there would be no distinction whatever between electrons, neutrinos, and quarks. If we lived at 10^{27} degrees Kelvin, the concept of grand unification would be commonplace, rather than novel.

A temperature of 10^{27} degrees Kelvin is, of course, outrageously large, even by the standards of astrophysics. The center of a hot star, for example, is only about 10^7 degrees Kelvin. The application of GUTs, however, forces us to consider such outrageous temperatures. Since the cosmological consequences of GUTs seem very attractive, one gains some confidence that our understanding of physics at these temperatures has a reasonable chance of being at least on the right track.

The baryon number of the universe

Having discussed the particle physics background, I will now return to the four questions that were introduced in Table 5.1. While these questions were left totally unanswered by the standard Big Bang model, I will now be able to show how the new ideas from particle physics can provide plausible answers to each of them.

I will begin with the first question, why the ratio of photons to protons and neutrons is about equal to 10^{10}. The idea that particle physics could provide an answer to this question was suggested first by Andrei D. Sakharov, and the more detailed calculation in the context of GUTs was first carried out by Motohiko Yoshimura of Tohoku University and by Weinberg. The study of this question was the first application of GUTs to cosmology, and the subject remains crucial to our understanding of cosmology in this context.

Particle physicists use the word 'baryon' to refer to either a proton or a neutron. More precisely, particle physicists define the 'baryon number' of a system by

$$\text{baryon number} \equiv \text{(number of protons)} + \text{(number of neutrons)}$$
$$- \text{(number of antiprotons)}$$
$$- \text{(number of antineutrons)} + \cdots,$$

where '$\cdots$' denotes the contributions from other particles which are very short-lived, and therefore irrelevant to the questions we are now discussing.

Obviously it is useful to have a single word to refer to either a proton or a neutron, because in the early universe the two types of particles rapidly interconvert, by processes such as

$$\text{proton} + \text{electron} \leftrightarrow \text{neutron} + \text{neutrino}.$$

The baryon number is unchanged by the reaction above, since the proton and the neutron each have a baryon number of one. In fact, all physical processes observed up to now obey the principle of baryon number conservation – the total baryon number of an isolated system cannot be changed. This principle implies, for example, that the proton must be absolutely stable; because it is the lightest baryon, it cannot decay into another particle without changing the total baryon number. Experimentally, the lifetime of the proton is now known to exceed 10^{32} years.

Note, however, that the principle of baryon number conservation does not forbid the production of baryons, provided that equal numbers of antibaryons are also produced. For example, at high energies the reaction

$$\text{electron} + \text{positron} \rightarrow \text{proton} + \text{antiproton}$$

is frequently observed.

To estimate the baryon number of the observed universe, one must ask whether the distant galaxies are composed of matter, or whether some might perhaps be formed from antimatter. This question has not been definitively answered, but there is a strong consensus that the universe is probably made entirely from matter. The belief is justified mainly by the absence of any known mechanism that could have separated the matter from the antimatter over such large distances. Assuming that this belief is true, then the total baryon number of the visible universe is about 10^{78}, corresponding to about one baryon per 10^{10} photons.

If the principle of baryon number conservation were absolutely valid, then the baryon number of the universe would be unchangeable. Under this assumption there would be no hope of explaining the baryon number of the observed universe – it would always have had a value of 10^{78}, a value that was necessarily fixed by the postulated initial conditions of the universe.

GUTs, however, imply that baryon number is *not* exactly conserved. At low temperatures, the conservation law is an excellent approximation, and the observed limit on the proton lifetime is consistent with at least many versions of GUTs. At temperatures of order 10^{27} degrees Kelvin and higher, however, processes that change the baryon number of a system of particles are expected to be quite common.

Thus, when GUTs and the Big Bang picture are combined, the net baryon number of the universe can be altered by baryon-nonconserving processes. However, to explain the observed baryon number, it is necessary that the underlying particle physics make a distinction between matter and antimatter. This distinction is essential, since any theory that leads to the production of matter and antimatter with equal probabilities would lead to a total baryon number far smaller than what is observed. For many years it had been thought that matter and antimatter behave identically, but a small difference between the two was discovered experimentally in 1964 by Val L. Fitch of Princeton University and James W. Cronin of the University of Chicago (who shared the 1980 Nobel Prize in Physics for this discovery). This inherent distinction between matter and antimatter has since been incorporated into many particle theories, including GUTs. Thus, in the context of GUTs, the observed excess of matter over antimatter can be produced naturally by elementary-particle interactions at temperatures near 10^{27} degrees Kelvin.

Finally, then, we come to the crucial question: Do GUTs give an accurate prediction for the baryon number of the observed universe? Unfortunately we cannot tell. The GUTs depend on too many unknown parameters to allow a quantitative prediction. However, one can say that the observed baryon number can be obtained with what seems to be a reasonable choice of values for these unknown parameters. Thus, GUTs provide at least a framework for answering the first question of Table 5.1. Sometime in the future, if the correct GUT and the values of its free parameters become known, it will be possible to make a real comparison between theory and observation.

The inflationary universe

The answers that I will discuss for the three remaining questions from Table 5.1 all depend on a new model for the very early behavior of the universe, a model called *the inflationary universe*. Before discussing the answers, I will have to spend some time describing how the model works. The description will be somewhat sketchy, but I will try to explain the main features.

The inflationary universe was first proposed by me in 1981, but the model in its original form did not quite work. It had a crucially important technical flaw, which was pointed out but not remedied in the original paper. A variation that avoids this flaw was invented independently by Andrei D. Linde of the P. N. Lebedev Physical Institute in Moscow and by Andreas Albrecht and Paul J. Steinhardt of the University of Pennsylvania. This chapter will discuss the Linde–Albrecht–Steinhardt version of the model, which is called 'the *new* inflationary universe.'

The key ingredient of the inflationary universe model is the assumed occurrence of a phase transition in the very early history of the universe. In Table 5.2 it was pointed out that GUTs imply that such a phase transition occurred when the temperature was about 10^{27} degrees Kelvin. This phase transition is linked to spontaneous symmetry breaking: At temperatures higher than 10^{27} degrees Kelvin, there is one unified type of interaction; at temperatures below 10^{27} degrees Kelvin, the grand unified symmetry is broken, and the $U(1)$, $SU(2)$, and $SU(3)$ interactions acquire their separate identities.

There are two possibilities for what might have happened when the universe cooled down to the temperature of this phase transition: The phase transition might have occurred immediately, or it may have been delayed, occurring only after a large amount of supercooling. (The word 'supercooling' refers to a situation in which a substance is cooled below the normal temperature of a phase transition, without the phase transition taking place. For example, water can be supercooled to more than 20 degrees Kelvin below its freezing point; and glasses are formed by rapidly supercooling liquids to temperatures well below their freezing points.) If the correct GUT and the values of its parameters were known, there would be no ambiguity about the nature of the phase transition – we would be able to calculate how quickly it would occur. In the absence of this knowledge, however, either of the two possibilities appears plausible. (Calculations do show, however, that only an extremely narrow range of parameters will lead to intermediate situations – in almost all cases the phase transition is either immediate or strongly delayed.)

If the phase transition occurred immediately, then its cosmological consequences would be very problematical. In that case a large number of exotic particles called magnetic monopoles would be produced, and the mass density of the universe would come to be strongly dominated by these particles. For most GUTs these monopoles would survive to the present day, leading to predictions which are grossly at odds with observation. Furthermore, in the case of an immediate phase transition, the last three questions of Table 5.1 would all remain unanswered.

The inflationary universe model is based on the other possibility, that the phase transition was delayed and the universe underwent extreme supercooling. As I will explain, the cosmological consequences of this assumption appear to be very attractive.

As the gas filling the universe supercooled to temperatures far below that of the phase transition, it would have approached a very peculiar state of matter known as a 'false vacuum.' This state of matter has never been observed. Furthermore, the energy density required to produce it is so enormous – about 60 orders of magnitude larger than the density of the atomic nucleus – that it clearly will not be observed in the foreseeable future. Nonetheless, from a theoretical point of view, the false vacuum seems to be well understood. The essential properties of the false vacuum depend only on the general features of the underlying particle theory, and not on any of the details. Even if GUTs turn out to be incorrect, it would still be quite likely that our theoretical understanding of the false vacuum would remain valid.

The false vacuum has a peculiar property which makes it very different from any ordinary material. For ordinary materials, whether they are gases, liquids, solids, or plasmas, the energy density is dominated by the rest energy of the particles of which the material is composed. (The rest energy is related to their masses by the famous Einstein relation, $E = mc^2$.) If the volume of an ordinary material is increased, then the density of particles would decrease, and therefore the energy density would also decrease. The false vacuum, on the other hand, is the state of lowest possible energy density that can be attained while remaining in the phase for which the grand unified symmetry is unbroken. This energy density is attributed not to particles, but rather to the Higgs fields which are responsible for the spontaneous symmetry breaking. Recall that we are assuming that the phase transition occurs very slowly; so, for a long time (by the standards of the very early universe) the false vacuum is the state with the least possible energy density that can be attained. Thus, even as the universe expands, the energy density of the false vacuum remains at a constant value.

When this peculiar property of the false vacuum is combined with Einstein's equations of general relativity, one finds a very dramatic result – the false vacuum leads to a gravitational repulsion. Throughout the rest of the universe's history gravity has slowed down the cosmic expansion; but, when the universe was caught in the false vacuum state, gravity actually caused the expansion to accelerate. This 'repulsion' is identical to the effect of Einstein's 'cosmological constant,' except that the repulsion caused by the false vacuum operates for only a limited period of time. Unfortunately, I have not found a way to explain the gravitational repulsion convincingly without invoking the detailed mathematics of general relativity, but in the next section I will describe in broad terms how the repulsion arises.

The gravitational repulsion would have produced a very rapid expansion, far in excess of the expansion rate of the standard Big Bang model. In the inflationary model essentially all the momentum of the Big Bang was produced by the gravitational repulsion. (In the standard Big Bang theory, by contrast, all the momentum of the Big Bang is incorporated into the postulated initial conditions.) The universe would double in size in about 10^{-34} seconds, and it would continue to double in size during each successive interval of 10^{-34} seconds for as long as the universe remained in the false vacuum state. During this period the universe expanded, or 'inflated,' by a stupendous factor. (A factor of at least 10^{75} (in volume) is necessary to answer the cosmological questions in Table 5.1, but the actual number depends on the highly uncertain details of the underlying particle theory, and may have been many orders of magnitude larger.)

Eventually, the phase transition would have to occur and, when it did, the energy density of the false vacuum would have been released. (In the language of thermodynamics, this energy is the 'latent heat' of the phase transition.) This energy input would have produced a vast number of particles and would have reheated the universe back to a temperature comparable to that of the phase transition: about 10^{27} degrees Kelvin. (The precise number is a factor of 2 or 3 below the temperature of the phase transition, but such factors go unnoticed in order-of-magnitude estimates.) The baryon-number-producing processes discussed in the previous section would have taken place during and just after the reheating – any baryon number density present before inflation would have been diluted to a negligible value by the enormous expansion. At the end of the phase transition, the universe would have been uniformly filled with a hot gas of particles, exactly as had been postulated as the initial condition for the standard Big Bang theory. Here, the inflationary model merges with the standard Big Bang theory; and the two models agree in their description of the evolution of the universe from this time onward.

In the inflationary model, virtually all the matter and energy in the universe were produced during the inflation. This seems strange, because it sounds like an unmistakable violation of the principle of energy conservation. How could it be possible that all the energy in the universe was produced as the system evolved?

In fact, the inflationary universe model is consistent with all the known laws of physics, including the conservation of energy. The loophole in the conservation of energy argument is due to the peculiar nature of gravitational energy. Using either general relativity or Newtonian gravity, one finds that *negative* energy is stored in the gravitational field. A simple way to make this fact seem at least plausible is to imagine two large masses, separated by a very large distance in an otherwise empty space. Now imagine bringing the two masses together. The masses will attract each other gravitationally, which means that energy can be extracted as the masses come together. One can imagine, for example, attaching the masses to fixed pulleys, with the wheels of the pulleys used to drive an electrical generator. Once the two masses are brought together, however, their gravitational fields will be superimposed, producing a much stronger gravitational field. Thus, the net result of this process is both to extract energy – and to produce a stronger gravitational field. If energy is conserved, then the energy in the gravitational field apparently goes down when its strength goes up. If the absence of a gravitational field corresponds to no energy, then any nonzero field strength must correspond to a negative energy. Gravitational energy is usually negligible under laboratory conditions, but cosmologically it can be very significant.

The energy stored in the false vacuum became larger and larger as the universe inflated, and was then released when the phase transition took place at the end of the inflationary period. At the same time, however, the energy stored in the cosmic gravitational field – the field by which everything in the universe is attracting everything else – became more and more negative. The total energy of the system was conserved, remaining constant at a value at or near zero.

Thus, inflation allows the entire observed universe to develop from almost nothing. The inflationary process could have started with an amount of energy equivalent to only about 10 kilograms of matter – even this small amount of energy could conceivably have been balanced by an equal contribution of negative energy in the gravitational field. Thus, if the inflationary model is correct, it is fair to say that the universe is the ultimate free lunch.

Negative pressure and gravitational repulsion

The mechanism that drives the accelerated expansion of the inflationary model cannot be described in detail without the formalism of general relativity, but in this section I will try to explain crudely how the gravitational repulsion arises. The material in this section is unnecessary for the rest of the chapter, so some readers may wish to skip immediately to the next section.

Recall that the false vacuum is the state of lowest possible energy density than can be attained while remaining in the phase for which the grand unified symmetry is unbroken. This energy density is attributed to the Higgs fields. In the vacuum state, the Higgs fields have nonzero values which break the grand unified symmetry, and in this state the energy density of the Higgs fields is zero, or at least very small. In the false vacuum, on the other hand, each Higgs field has a value of zero, preserving the grand unified symmetry. In order to achieve the spontaneous symmetry breaking, however, the theory was formulated so that the state in which all the Higgs fields vanish is a state of high energy density! (Although it seems strange that energy should be required in order for the value of the Higgs fields to be zero, particle physicists find that this property causes no inconsistencies, and is exactly what is needed to produce the spontaneous symmetry breaking.) Thus, as a region of false vacuum expands, the energy density remains constant – it is just the energy density necessary to maintain a value of zero for the Higgs fields.

The constancy of the energy density of the false vacuum is related to another very peculiar property – the false vacuum has a pressure which is both large and negative. To understand the connection between these two properties, consider the fact that when a normal, positive pressure gas is allowed to expand, it will push on its surroundings and in the process it will lose energy to its surroundings. Both steam and gasoline engines operate on this principle. For the false vacuum, however, the situation is reversed. We can imagine a region of false vacuum that expands, but the expansion occurs at a constant energy density. The energy of the false vacuum region therefore increases as the volume increases, which means that the region is taking energy from its surroundings. This indicates that the region must create a negative pressure, or suction, so that energy is being supplied by whatever force is causing the expansion. By considering the energy balance involved in the expansion of a region of false vacuum, it is possible to determine the pressure uniquely – the pressure is equal to the negative of the energy density, when the two are measured in the same units.

According to Newton's theory of gravity, a gravitational field is produced by a mass density. In a relativistic theory the mass density can be related to a corresponding energy density by $E = mc^2$. According to Einstein's theory of general relativity, however, a pressure can also produce a gravitational field. When Einstein's equations are used to describe a homogeneously expanding universe, they show that the rate at which the expansion is slowed down is proportional to the energy density plus three times the pressure. Under ordinary circumstances the pressure term is a small relativistic correction, but for the false vacuum the pressure term overwhelms the energy-density term and has the opposite sign. So the bizarre notion of negative pressure leads to the even more bizarre effect of a gravitational force that is effectively repulsive.

Answers to remaining questions

Having described the foundations of the inflationary universe model, I can now explain how the remaining questions of Table 5.1 can be resolved. First, I will discuss Question #2, concerning the large-scale homogeneity of the universe. Recall that in the standard Big Bang theory, the large-scale homogeneity cannot be explained because the universe did not have enough time to come to a uniform temperature.

Consider now the evolution of the observed region of the universe, which has a radius today of about 10 billion light years. Imagine following this region backward in time, using the inflationary model. Follow it back to the instant immediately before the inflationary period. Since the theory predicts a tremendous spurt of expansion during the inflationary period, one infers that the region was incredibly small before this expansion began. In fact, the region was more than a billion times smaller than the size of a proton. (Note that I am *not* saying that that universe as a whole was very small. The inflationary model makes no statement about the size of the universe as a whole, which might in fact be infinite.)

While the region was this small, there was plenty of time for it to have come to a uniform temperature. So in the inflationary model, the uniform temperature was established before inflation took place, in a very, very small region. The process of inflation then stretched this very small region to become large enough to encompass the entire observed universe. Thus, the sources of the microwave background radiation arriving today from all directions in the sky were once in close contact; they had time to reach a common temperature before the inflationary era began.

The inflationary model also provides a simple resolution for Question #3, the issue of the mass density. Recall that the ratio of the actual mass density to the critical density is called Ω, and that the problem arose because the condition $\Omega = 1$ is unstable: Ω is always driven away from 1 as the universe evolves, making it difficult to understand how its value today can be in the vicinity of 1.

During the inflationary era, however, the peculiar nature of the false vacuum state results in some important sign changes in the equations that describe the evolution of the universe. During this period, as we have seen, the force of gravity acts to accelerate the expansion of the universe rather than to retard it. It turns out that the equation governing the evolution of Ω also has a crucial change of sign: During the inflationary period the universe is driven very quickly and very powerfully *towards* a critical mass density.

In other words, a very short period of inflation can drive the value of Ω very accurately to 1, no matter where it starts out. There is no longer any need to assume that the initial value of Ω was incredibly close to 1.

Furthermore, there is a prediction that comes out of this. The mechanism that drives Ω to 1 almost always overshoots, which means that even today the mass density should be equal to the critical mass density to a high degree of accuracy. More precisely, the model predicts that the value of Ω today should equal 1 to an accuracy of about one part in 10 000. (The deviations from 1 are caused by quantum effects, which I will talk about shortly.) Thus, the determination of the mass density of the universe would be a very important test of the inflationary model.†

† In the text I have followed the common assumption that Einstein's cosmological constant Λ is either zero or negligible. Otherwise the prediction becomes $\Omega + (\Lambda/3H^2) = 1$.

Unfortunately, it is very difficult to estimate the mass density of the universe reliably. Part of the reason is that most of the mass in the universe is in the form of 'dark matter,' matter that is totally unobserved except for its gravitational effects on other forms of matter. Since we do not even know what the dark matter is, it is very difficult to estimate how much exists. Most of the current estimates, I must admit, give values for Ω that are distinctly below 1: Numbers like 0.1–0.3 are most common. But these estimates are highly uncertain, and there appears to be no compelling observational evidence at present to rule out the possibility that $\Omega = 1$.

Finally, then, I come to the last of the four questions, concerning the origin of the primordial density perturbations in the universe. The generation of density perturbations in the new inflationary universe was addressed in the summer of 1982 at the Nuffield Workshop on the Very Early Universe, held at Cambridge University. A number of theorists were working on this problem, including Steinhardt, James M. Bardeen of the University of Washington, Stephen W. Hawking of Cambridge University, So-Young Pi of Boston University, Michael S. Turner of the University of Chicago, A. A. Starobinsky of the L. D. Landau Institute of Theoretical Physics in Moscow, and me. We found that the new inflationary model, unlike any previous cosmological model, leads to a definite prediction for the spectrum of perturbations. Basically the process of inflation first smoothes out any primordial inhomogeneities that may have been present in the initial conditions. For example, any particles that may have been present before inflation are diluted to a negligible density. In addition, the primordial universe may have contained inhomogeneities in the gravitational field, which are described in general relativity in terms of bends and folds in the structure of spacetime. Inflation, however, stretches these bends and folds until they become imperceptible, just as the curvature of the surface of the Earth is imperceptible in our everyday lives.

For a while, we were worried that inflation would give us a totally smooth universe, which would be obviously incompatible with observation. It was pointed out, however, I believe first by Hawking, that the situation might be saved by the application of quantum theory.

A very important property of quantum physics is that nothing is determined exactly – everything is probabilistic. Physicists are, of course, accustomed to the idea that quantum theory, with its probabilistic predictions, is essential to describe phenomena on the scale of atoms and molecules. On the scale of galaxies or clusters of galaxies, however, there is usually no need to consider the effects of quantum theory. But inflationary cosmology implies that for a short period the scales of distance increased very rapidly with time. Thus, the quantum effects which occurred on very small, particle physics length scales were later stretched to the scales of galaxies and clusters of galaxies by the process of inflation.

Therefore, even though inflation would predict a completely uniform mass density by the rules of classical physics, the inherent probabilistic nature of quantum theory gives rise to small perturbations in the otherwise uniform mass density. The spectrum of these perturbations was first calculated during the exciting 3-week period of the Nuffield Workshop. After much disagreement and discussion, the various working groups came to an agreement on the answer. I will describe these results in two parts.

First of all, we calculated the shape of the spectrum of the perturbations. The concept of a spectrum of density perturbations may seem a bit foreign, but the analogy of sound waves is very close. People familiar with acoustics understand that no matter how complicated a sound wave is, it is always possible to break it up into components which each have a standard wave form and a well-defined wavelength. The spectrum of the sound wave is thus specified by the strength of each of these components. In discussing density perturbations in the universe, it is similarly useful to define a spectrum by breaking up the perturbations into components of well-defined wavelength.

For the inflationary model, we found that the predicted shape for the spectrum of density perturbations is essentially scale-invariant; that is, the magnitude of the perturbations is approximately equal on all length scales of astrophysical significance. While the precise shape of the spectrum depends on the details of the underlying GUT, the approximate scale-invariance holds in almost all cases. The scale-invariant spectrum is in agreement with a phenomenological model for galaxy formation proposed in the early 1970s by Edward R. Harrison of the University of Massachusetts at Amherst and Yakov B. Zel'dovich of the Institute of Physical Problems in Moscow, working independently.

Unfortunately, there is still no way of inferring the precise form of the primordial spectrum from observations, since one cannot reliably calculate how the universe evolved from the early period to the present. Such a calculation is very difficult in any case, and it is further complicated by the uncertainties about the nature of the dark matter. Nonetheless, the scale-invariant spectrum appears to be, at least approximately, what is needed to explain the evolution of galaxies, and thus this prediction of the inflationary model appears so far to be successful. Galaxy formation is currently a very active subject of research, so a better determination of the spectrum of primordial density perturbations may be developed. Such a result would provide an additional test of the inflationary universe model.

The predicted magnitude of the density perturbations was also calculated by the group at the Nuffield Workshop, but the implications of these results were much less clear. It was found that the predicted magnitude, unlike the shape of the spectrum, is very sensitive to the details of the underlying particle theory. At the time, the minimal $SU(5)$ theory – the first and simplest of the GUTs – was strongly favored by anybody interested in GUTs. We were therefore very disappointed when we found that the minimal $SU(5)$ theory leads to density perturbations with a magnitude 100 000 times larger than what is desired for the evolution of galaxies. Thus, there was a serious incompatibility between the inflationary model and the simplest of the GUTs.

With the passage of time, however, the credibility of the minimal $SU(5)$ grand unified theory has diminished. The minimal $SU(5)$ theory makes a rather definite prediction for the lifetime of a proton, and a variety of experiments have been set up to test this prediction by looking for proton decay. So far, no such decays have been observed, and the experiments have pushed the limit on the proton lifetime to the point where the minimal $SU(5)$ theory is now excluded.

With the exclusion of the minimal $SU(5)$ theory, a wide range of GUTs become plausible. All of the allowed theories seem a bit complicated, so apparently we will need some kind of new understanding to choose which – if any – is correct.

A variety of GUTs that predict an acceptable magnitude for both the proton lifetime and the density perturbations have been constructed. Thus, while the inflationary model cannot be credited with correctly predicting the magnitude of the perturbations, it also cannot be criticized for making a wrong prediction. The situation is very similar to the calculation of the net baryon number of the universe which I discussed earlier: The inflationary model provides at least a framework for calculating the magnitude of the density perturbations. If sometime in the future the correct GUT and the values of its free parameters somehow become known, it will then be possible to make a real theoretical prediction for the magnitude of the perturbations.

A common feature of those models leading to acceptable density perturbations is the abandonment of the idea that inflation can be driven by the Higgs fields that break the grand unified symmetry. It appears that any Higgs field that interacts strongly enough to break the grand unified symmetry leads to density perturbations with a magnitude that is far too large. Thus it must be assumed that the underlying particle theory contains a new field – a field which strongly resembles the Higgs fields in its properties, but which interacts much more weakly.

Unfortunately, all of the known theories that give acceptable predictions for the magnitude of the density perturbations look a little contrived. Well, to be honest, the theories *were* contrived – with the goal of getting the density perturbations to come out right. The need for this contrivance can certainly be used as an argument against the inflationary model; but, in my opinion, this argument is considerably weaker than the arguments in favor of inflation. Even if we ignore cosmology, any GUT that is consistent with the known properties of particle physics appears to be rather contrived. Clearly there are some fundamental principles at work here that we do not yet understand.

I would like to emphasize that my allusions to fundamental principles beyond GUTs are not based on idle speculation – they are based on active and energetic speculation. Since the invention of GUTs in 1974, particle theorists have been vigorously working on attempts to construct the ultimate theory of nature – an elegant theory which would include a quantum description of gravity. The characteristic energy scale of such a theory is presumably the Planck scale, 10^{19} GeV, a point at which the gravitational interactions of elementary particles become comparable in strength to the other types of interactions. It is then hoped that a GUT would emerge as a low-energy approximation.

The latest and most successful of these attempts at unification is a radically new kind of particle theory known as 'superstring theory'. Superstrings represent a dramatic departure from conventional theories in that particles are viewed as ultramicroscopic strings (length $\approx 10^{-33}$ centimeters). Furthermore, according to this theory, the universe has *nine* spatial dimensions. Early in the history of the universe, when the temperature cooled below 10^{32} degrees Kelvin, all spatial dimensions, except the three we know today, stopped expanding and remained curled up with an unobservably small extent. As bizarre as the theory may sound, the superstring theory has been shown to possess a number of unique properties crucial to a quantum theory of gravity, and it has totally captured the attention of a large fraction of the worldwide particle theory community.

For now, very little can be said about the behavior of superstring theories at energies well below the Planck scale. Nonetheless, it is encouraging to know that progress is under way toward embedding the idea of grand unification into a larger framework. Superstring theories are highly constrained, which leads to hopes that someday we may be able to make rather definite predictions concerning physics at the energy scale of GUTs and beyond. If such a success is ever achieved, then the calculation of the predicted spectrum of density perturbations will provide a very rigorous test of inflationary cosmology.

Finally, I want to mention that quantum effects during the inflationary era are not the only potential source of primordial density perturbations. There is also the possibility, which I will not discuss in detail, that the seeds for galaxy formation may have been objects called 'cosmic strings' (not related to 'superstrings'). These strings are predicted by some (but not all) GUTs, and they would have formed in a random pattern during the GUT phase transition. As their name suggests, strings are very thin, spaghetti-like objects that can form infinite curves or closed loops of astrophysical size. With a thickness of about 10^{-29} centimeters, a cosmic string has a mass of about 10^{22} grams for each centimeter of length. (In astronomical terms that is equal to about 10^7 solar masses per light year.) In most theories, the density of these strings would be diluted to negligibility by the process of inflation. However, it is possible to construct theories in which the strings survive by forming either after inflation or at the very end of it. Cosmic strings are a very active topic of current research, particularly since they can explain naturally a number of features of galactic structure. Models of this type still make use of inflation to answer Questions #2 and #3 of Table 5.1, and also to smooth out any small-scale inhomogeneities which may have been present in the initial conditions.

Conclusion

In summary, the inflationary universe model has been very successful in describing the broad, qualitative properties of the universe. In particular, the model provides very attractive answers to the four questions discussed in this chapter. While the model must be treated as speculative, I nonetheless feel that in its broad outline the concept of an inflationary universe is essentially correct.

The inflationary model makes two observationally testable predictions – it predicts the mass density of the universe, and also the shape of the spectrum of primordial density perturbations. While neither of these predictions is straightforward to check, it seems likely that significant progress will be made in the foreseeable future.

Even if the inflationary model is correct, however, I must still emphasize that nothing discussed here is a completed project. The inflationary model is not a detailed theory – it is really just an outline for a theory – Michael Turner has called it the 'inflationary paradigm.' To fill in the details, we will need to know much more about the properties of particle physics at the energy scales of GUTs – and perhaps beyond.

It looks to me as if the fields of particle physics and cosmology will remain closely linked for some years to come, as physicists and astronomers continue their efforts to understand the fabric of space, the structure of matter, and the origin of it all.

Further reading

Blau, S. K. and Guth, A. H. Inflationary cosmology. *300 Years of Gravitation* (Editors S. W. Hawking and W. Israel), Cambridge: Cambridge University Press, 1987.

Dicke, R. H. and Peebles, P. J. E. The big bang cosmology – enigmas and nostrums. *General Relativity: An Einstein*

Centenary Survey (Editors S. W. Hawking and W. Israel), Cambridge: Cambridge University Press, 1979.

Gribbin, J. *In Search of the Big Bang: Quantum Physics and Cosmology*, Heinemann: London, 1986.

Guth, A. H. and Steinhardt, P. J. The Inflationary Universe. *The New Physics* (Editor Paul Davies), Cambridge: Cambridge University Press, in press.

Hey, T. and Walters, P. *The Quantum Universe*, Cambridge: Cambridge University Press, 1987.

Schramm, D. N. The early universe and high-energy physics. *Physics Today*, April 1983.

Silk, J. *The Big Bang: The Creation and Evolution of the Universe*, New York: W. H. Freeman, 1980.

Tyron, E. P. Cosmic inflation. *The Encyclopedia of Physical Science and Technology*, New York: Academic Press, 1987.

Weinberg, S. *The First Three Minutes: A Modern View of the Origin of the Universe*, New York: Basic Books, 1977.

6

Expanding the universe: Space telescope and beyond in the next twenty years

JAMES E. GUNN
Princeton University

Introduction: astronomy as big science

In the modern age, the advance of astronomy has been led by developments in instrumentation, beginning with the precise measuring instruments of Tycho Brahe. The data provided by Tycho allowed Kepler to formulate his laws of planetary motion, which, in turn, led Newton to the laws of gravitation. At much the same time, Galileo's first telescopes opened a vista onto a visible universe of enormous complexity.

Tycho's efforts were already big science in the most pejorative sense of that term. Without royal support from the Danish king and essentially the whole economic output of a sizeable population on the island of Hveen, Tycho could not have done what he did. Galileo's efforts were more modest (although certainly not without political import); but, the demand for larger and larger telescopes quickly made observational astronomy impossible without support from governments or, at least, the very rich. It would be well for those who bemoan the current dependency of astronomical research on governmental support to consider the history of the subject. Still, the bemoaners have many valid points, as we will see anon.

In any case, we are now talking about very large and expensive endeavors if we wish to improve substantially upon existing knowledge. The fact carries with it penalties and responsibilities, a subject which we will also have more to say later.

Any discussion of instruments is nonsensical without some understanding of the problems they are designed to address. Let us therefore consider some of the outstanding cosmological questions of the day and how we have learned as much (or as little) about them as we have done.

Six questions

A crucial aspect of any physical science not well appreciated by those unschooled in its methods is that much more important than answering questions is the art of asking the right ones. The progress of science is mostly marked by realizations that wrong, or inappropriate, questions are being asked, and the undeniable impact of genius is most felt in that quarter. While genius has also contributed to solutions of problems, those solutions probably would have come anyway by dint of hard labor. But the questions themselves almost certainly would not have come by that route.

I pose here six questions central to current astrophysical research. All have been discussed by other authors in this volume, but I will use them here as a focus for the potential instrumental advances which form my main theme.

(1) How big is the universe?

This is one of those questions which has changed dramatically over the last few years in response to a revolution in our thinking about the origin of the universe. It has two parts: a very practical (but difficult) one and a more philosophical one, which may or may not be a good question. Instrumentation has direct bearing on the first part: What is the distance scale in the universe – that is, how far away are the galaxies we can see? In the late twenties and early thirties, it became clear that the universe was *expanding*, in that distant galaxies were moving away from us at speeds which were in direct proportion to their distances. The distances to the galaxies were (and still are) not known well, but there have come to be better and better ways to estimate the *relative* distances. The expansion is succinctly stated in *Hubble's Law*: Velocity = Hubble's Constant × distance, with the velocities of galaxies measured directly by means of their Doppler redshifts. Unfortunately, Hubble's Law is only an approximation to the truth; in addition to the expansion velocity, galaxies have random velocities of their own, induced probably by gravitational accelerations from their neighbors. Only at fairly large distances do the expansion velocities completely dominate the random ones, and that fact is the crux of the tale.

For very nearby galaxies like our nearest large neighbor, the Andromeda Spiral, about 2 million light years distant, we can study individual stars just as we study those in our own galaxy. Because we know the distances to such stars in our own system with fair accuracy and thus know their intrinsic brightness, we can use this knowledge to infer the distances to those galaxies in which similar stars can be seen. But galaxies measured by these techniques have velocities completely dominated by local gravitational interaction. Indeed, the Milky Way and Andromeda are *approaching* one another due to their mutual attraction; 5 billion years or so hence, they will be very close together and may even collide. In short, there are no galaxies for which we have both good distances *and* velocities which are accurate reflections of the expansion of the universe. This means we have no way to determine directly the value of the Hubble Constant, which sets the scale of the universe, by simply taking the ratio of two accurately known quantities, distance and velocity. Instead, we must seek distances by a tortuous chain of secondary indicators, with the result that we know Hubble's Constant with an accuracy of only about 50 percent. As we shall see, this affects not only our notion of the universe's size but also its age and its physical evolution at early times. It is an entirely intolerable state of affairs.

How do we get around the problem? The straightforward approach seems to be to enlarge the number of galaxies in which we can study stars in detail – either to distances large enough to trust the velocities, or at least to distances large enough to calibrate reliably the best of the secondary distance indicators. The key to this is improved *angular resolution*. From the ground, atmospheric turbulence limits the resolution of pictures of astronomical objects to about one arcsecond, 1/3600 of a degree. This blurring makes it impossible to resolve, or separate, individual stars in any but the very nearest galaxies. For example, Figure 6.1 shows a galaxy about fifteen times as distant as Andromeda. Although this photograph was taken under excellent conditions from the ground, no individual stars can be resolved. Only a large and very good telescope in space can do that trick. This, indeed, is a primary reason for the Hubble Space Telescope (HST). Figure 6.2 shows the

Figure 6.1 A galaxy (NGC 7331) similar to the Andromeda Spiral, but about fifteen times more distant. Even with the largest telescopes and best observing conditions, no individual stars can be resolved. Images taken with the HST will show this object in similar detail to Figure 6.2. (Hale Observatories photograph.)

outer part of the Andromeda Galaxy in a photograph from the ground; with HST we will be able to get comparable images for galaxies more than ten times as distant. Such resolution offers considerable hope that HST will help determine the value of the Hubble Constant to an accuracy of a few percent.

The larger, almost philosophical, question is: 'How big is the universe *really*?' Actually, the issue of whether the universe is finite or infinite has, in the last 10 years, fallen from one about which workers looked forward to an answer in their lifetimes to one which – if it has any answer at all – most probably cannot be determined by any measurement we can ever make. This dramatic change in view has been brought about by the equally dramatic idea of inflation, discussed in the preceding chapter by one of its inventors, Alan Guth. Although fascinating, it clearly is not germane to our discussion of instruments, and we will leave it.

Figure 6.2 The outer parts of the Andromeda Spiral, the nearest large neighbor galaxy to our own. It is about 2 million light years distant. Individual stars making up the spiral arms can be easily seen, and the brightest can be studied in some detail. This is to be contrasted to Figure 6.1. (Photograph © by Mount Wilson Observatory/Carnegie Institution of Washington.)

(2) How old is the universe?

Whether this question is even meaningful has historically been a matter of some debate. However, today, the consensus is that the universe began in a cataclysmic event called the Big Bang, which occurred at some finite time in the past. Indeed, the existence of the Big Bang is strongly suggested by Hubble's Law itself. Since the recession velocity of galaxies is proportional to their distances, and if we assume velocities do not change with time, then any observed galaxy at some known distance (d), with some known velocity (V), would at a time in the past which is determined by the ratio of this distance to this speed (d/V) have been right here, on top of us. Further, since any galaxy twice as far away is traveling twice as fast, its indicated time of departure would also be *the same* – and so it is for any galaxy at any distance. This time – or starting point – is just the reciprocal of Hubble's Constant, so measuring the constant clearly also has something to do with measuring the age of the universe.

In fact, the supposition that velocities do not change with time is probably incorrect. Since gravity operates on all matter in the universe, the attractive force slows expansion. Thus, the universe was expanding more rapidly in the past. (Hubble's 'Constant' may change with time, but the *form* of the law is always the same.) A more rapid expansion in the past means that the universe must be younger than the age calculated without gravitation (the 'Hubble age') by a figure dependent on how much deceleration there has been. To calculate this deceleration, one must know how much matter there is, a problem we shall come to later, but it is probable that the real age is somewhere between the 'calculated Hubble age' and an age half that amount. On the other hand, if we live in an inflationary universe, the real age is probably exactly 2/3 the Hubble age.

Unfortunately, we do not know the Hubble age with any certainty. Values between 10 and 20 billion years can be – and are – defended.

A necessary complement to the Hubble age is the measurement of the ages of *things* in the universe. It would be embarrassing to the extreme if an indisputable age for a star, for instance, were greater than the age of the universe derived from the expansion. Things can be younger than the universe, but nothing better be older. Indeed, there are structures now believed to be very old, having formed no later than a billion years after the Big Bang. An unambiguous determination of their ages would be a strong constraint on possible models of the universe. But again, we need to know the Hubble age better.

Among the potpourri of things whose ages might be measured, the most promising seem to be naturally radioactive chemical elements with long half-lives – and stars. The evidence from both so far suggests the universe is old – older than 10, probably about 15 billion years, and certainly more than the shorter Hubble ages derived from a decrease due to deceleration. In this area we expect no breakthroughs due to new instrumentation. What is necessary is painstaking attention to detail, to making ever more accurate models of stars, and to understanding the complex geochemistry of the radioactive elements both in the Earth and in meteorites. Of course, all that effort is useless as a test for cosmological ages without a better Hubble Constant, so we come again to Space Telescope.

(3) Will the universe expand forever?

Many physicists regard this as yesterday's question. If inflation is correct, the answer is yes. More accurately, the answer is that the expansion will continue for so long that it *really* doesn't matter. By the time the universe shows signs of leaning one way or another, the part we can survey will have grown so large that our present horizon will encompass only an infinitesimal piece of it. Certainly, what the universe will look like then cannot be predicted by *any* measurement we can make now.

The classical concern with this question was motivated by the notion that the observable universe is a fair sample of the whole. Thus, the question was basically a simple one: Is the universe expanding faster than its local escape velocity – or not? Any sphere in the universe that contains some amount of mass has a gravitational field associated with an escape velocity, just as the escape velocity from the surface of the Earth is 7 kilometers per second. Since doubling the radius of a sphere also doubles the escape velocity (if both the larger and the smaller are large enough to take in a fair average of the mass density in the universe), the Hubble expansion velocity will always be greater or less than the escape velocity by the same factor. If the Hubble velocity is less than the escape velocity, gravity will win the battle. Expansion will slow and stop, and the universe will fall together again to end in a cataclysmic crunch, just as it began in a cataclysmic bang. If, on the other hand, the velocity is greater than the escape velocity, each galaxy will escape every other, and the expansion will go on forever into an ever lonelier future.

How do we find the answer? Ideally, we'd like to measure the rate of expansion in the past and *observe* how much (and, indeed, whether) it has slowed down since. This involves measuring the velocities and distances of very distant objects, so distant that the light we see left them at some time reasonably early in the universe's history. Fortunately, we need only relative distances, so the Hubble Constant fiasco is not of immediate concern. But now we have another problem: the objects we are looking at are seen in their youth, and are doubtless different from their nearby counterparts seen in old age, so we must know how they *evolve*. If they were brighter in the past, we might even think they are closer than they really are. To study their evolution requires studying their spectra in great detail to determine what stars are present and what those stars are made of. To do that for faint, distant objects requires really big telescopes. The 200-inch (5-meter) Hale telescope on Mount Palomar has made a start, and the 10-meter Keck telescope under construction by Caltech and the University of California on Mauna Kea push it farther, but it is unclear whether even it is large enough. The Space Telescope will give us clear pictures of the structure of these distant objects, but it is far too small to reveal the details of their stellar populations.

(4) What is the universe made of?

Another approach to determine the expansion rate is to measure the amount of matter in the universe 'locally,' say, in the volume out to some 300 million light years. From this, one could *calculate* the gravitational deceleration. However, attempts to do this, as well as evidence from many other lines of endeavor, have revealed the dismaying fact, as discussed earlier by Vera Rubin, that most of the universe's mass must be made of some dark, completely unknown matter. There is some evidence, and it is indeed the current fashion to believe, that the dark matter is not ordinary matter made of protons and neutrons and electrons, but rather some new, stable, exotic neutral elementary particle. We know a great deal about the *behavior* of the dark matter, but we have no observational clues as to what it may be, and only vague theoretical ideas about the possibilities. (Of course, there is no dearth of theoretical candidates; theorists are very inventive.)

It is unlikely that astronomical observations will shed much light on this problem. Considerable ingenuity is being expended now on devices which will detect the dark matter directly if the particles interact with ordinary matter through the weak nuclear force, as is expected for some candidates. The particle itself might possibly be produced in very big accelerators, or, at least, theories which predict the existence of the most promising candidates can be tested in the relevant energy ranges. Somehow, it is not entirely satisfactory to live in a universe dominated by something of completely unknown nature. Here, perhaps, we are waiting for genius.

(5) How did the structure we see develop?

By this question, we generally mean 'How did galaxies and clusters of galaxies develop?' But, unless we want to ask seven questions, we must add 'How do stars form?' The question of the development of large-scale structure, that is, the distribution of galaxies *vis-a-vis* their neighbors, is being hotly pursued at the present, at least partly because it is related to understanding the nature of the dark matter. While the effort is basically a cataloging one, it is an area in which observations make very direct contributions.

The picture of galaxy distribution nearby is slowly yielding to painstaking efforts like the one Margaret Geller has described, but many years of work still lie ahead. These projects require a body of very dedicated workers, enormous amounts of telescope time and very sensitive detectors, but (fortunately) not the very largest telescopes. All of the requirements seem to be met, and the work is proceeding as rapidly as one might hope. For example, the new Palomar Sky Survey will be a gold mine of data, if the money can be found to analyze it properly. (One of the infinitely sad things about astronomy – and, for that matter, many other sciences – is that money for flashy measurements or glamorous space missions is not too difficult to come by, but money to analyze the results is very difficult to get. Since it is the analysis that produces *knowledge*, one only hopes that the funding agencies will react sanely in this case.)

Very large telescopes are required, however, if we are to chart the development of structure by measuring the distribution of galaxies at very great distances and, hence, very early times. Similar instruments are needed to investigate the structure of galaxies themselves in their youth. The former effort requires redshifts of thousands, or better, tens of thousands, of distant galaxies. One recent technique making use of new optical designs for spectrographs allows the simultaneous recording of the spectra of many objects in the same field. Although a 'dedicated' instrument might profitably do nothing else, there is simply not enough time available on the largest general-purpose telescopes to mount such an effort today. The effort of looking directly at the evolution of the forms of galaxies goes hand in hand with studying their spectral evolution, that is, studying their population of component stars. The former requires all the imaging power of the HST, the latter all the collecting area of the largest ground-based telescopes. The result, to see cosmic history and evolution unfolding before our eyes, is to me the most exciting prospect for astronomy in the future.

Finally, stars. While we understand a fair amount (but certainly not everything) about how stars evolve and die, we know little about how they form. Observationally, the problem is very difficult: Stars form deep in dense clouds of gas and dust, and these wombs are impenetrable to optical and near-infrared radiation. Only in the far-infrared wave band, which is very difficult to observe from the ground, and in the radio wave band, which does not arise directly from the protostar, does radiation emerge from the region of formation. Nothing will be built soon in the United States to address this problem, although some instruments proposed for the future and several programs in Europe and Japan are addressing the problem. No one dreams that stars have anything to do with the exotic processes occurring during the early universe. Still, the birth of stars is surrounded by, if anything, more mystery than the formation of galaxies. Since we see galaxies by the light of the stars of which they are made, we cannot really say we understand how galaxies – or the universe – evolved unless we understand how stars form.

(6) What is the nature of strong gravitational fields?

Gravity is even more inexorable than death and taxes. A star with the mass of the Sun, once all its thermonuclear resources are spent, will be collapsed by gravity into a sphere less than 1 percent of its present size, with a density in its center of 1000 tons per cubic inch. A star a little less massive than twice the mass of the Sun cannot stop there; gravity carries it on to become a neutron star, a body about 10 miles in diameter with densities approximately like those in the nuclei of atoms, roughly a billion tons per cubic inch. For bodies a little more massive yet, there is no stable end point. In fact, we believe now that a black hole, an object whose gravity field is so strong not even light can escape, inevitably must form, and that deep in its interior lurks a real physical singularity in which the density is infinite. It is easy to imagine that massive objects might collect in the centers of galaxies and collapse to form a black hole. Indeed, there is abundant evidence that some such thing is going on – even in the Milky Way.

Although no radiation can escape from the black hole proper, its immense field attracts gas from its environs on which it feeds and grows. That gas probably forms a disk in orbit about the hole. Since gas close to the hole orbits faster than gas farther out, there is friction in such a disk, and that friction generates very high temperatures. Hence, the stuff surrounding a black hole can be very luminous indeed. Some of the radiation is swallowed by the hole as the disk slowly feeds into the waiting monster, but most escapes. This is one explanation for quasars, the most luminous objects in the universe and among the most mysterious.

We know very little more about quasars than we knew when they were discovered some 25 years ago. The description above is a simplified summary of what most people believe; however, in truth, there is not a shred of direct evidence for such a picture. Their energy output is so enormous that gravity is universally believed to be the ultimate culprit, but no one has actually seen what happens in the environs of a black hole. (On the other hand, there *is* very good evidence that this kind of phenomenon occurs near black holes which are the binary companions of ordinary stars. And, since a number of these have been detected, the beasts almost certainly exist.) To study the phenomenon, one must have angular resolution very much better than even what is possible with Space Telescope. Most of the activity associated with a 100-million-solar-mass black hole – the size needed to explain the quasar phenomenon – is contained within a region only about the size of our solar system. In the Virgo Cluster, the nearest large cluster of galaxies suspected of containing such an object, a region this size would be 1/10 000 of an arcsecond as seen from the Earth. An optical telescope large enough to resolve it would have to be 2000 meters in diameter! An instrument now under construction by the National Science Foundation may help: the Very Long Baseline Array (VLBA). This array of radio telescopes can act together to form a telescope almost as large as the Earth. Radio waves are much longer than optical ones, so telescopes need to be proportionally larger, but the VLBA working at its short-wavelength limit will come close to the required resolution to study the very nearest examples.

There are other ways to gain invaluable but indirect information. The very strong electromagnetic fields associated with quasars are very efficient particle accelerators, putting the best of our puny efforts to shame. It is, in fact, these fast particles which give rise to the radio (and perhaps the optical) radiation from these objects. The particles interact with both the radiation and matter to make X and gamma rays. Quasars are seen in most cases to be X-ray sources and, in a few cases, gamma-ray sources. Even though X-ray telescopes lack high resolution (the best to date is roughly comparable to ground-based optical telescopes, or about 1 arcsecond), the X rays probably come from deep in the quasar, very near the black hole, and almost certainly carry information about conditions there, could we only interpret it.

In very many cases, the structure around the object may direct the energetic particles in one or two narrow jets which carry energy incredibly efficiently to regions far from the galaxy, where it is finally deposited and radiated away, mostly as radio emissions. These double radio sources are common, more common by far than quasars, and suggest that while the black hole engine itself may not always be bright in either the radio or the optical wavebands, it can still far out-accelerate the most ambitious superconducting supercollider ever dreamt of by Earthly physicists.

It is safe to say that these objects will not yield their secrets easily, yet the pursuit is one of the most exciting ever undertaken by science.

Instruments on the ground: present and future

The largest productive optical telescope today is the 200-inch Hale telescope on Mount Palomar (Figure 6.3). Designed in the 1930s, it came into regular operation about 1950. For many years, most large telescopes have been built very much like the Hale instrument, but few out-perform it in any significant way. Even though the collecting area of optical telescopes is no larger than it was 35 years ago, that does not mean the power of such instruments has not grown. Indeed, it has done so dramatically through improvements in optics and in the detectors attached to them.

Twenty years ago, the primary detector for almost all imaging applications in cameras and in spectrographs was the photographic plate. In that era, photographic plates at best had *quantum efficiencies* of only about 0.1 percent, that is, only one in a thousand optical photons, the quantum particle which makes up light, was actually recorded. Because the time required to make a given measurement to a given accuracy is inversely proportional to the quantum efficiency, it took a thousand times longer to make a given observation than it would have with a 'perfect' detector. We still do not have perfect detectors, but the solid-state cameras called CCDs now come very

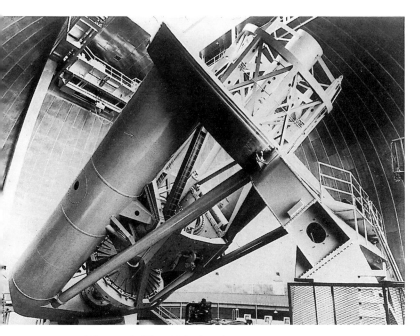

Figure 6.3 The 200-inch Hale telescope on Mount Palomar, currently the largest productive telescope in the world. Recent advances in instrumentation have made it thousands of times more sensitive than it was when built, but fundamental limits have now very nearly been reached, and more sensitive instruments in the future will have to be larger. (Hale Observatories photograph.)

close. The quantum efficiencies of CCDs over broad wavelength regions in the visible and near infrared is about 70 percent. Combined with improvements in optical coatings and design, today's instruments can have *system* quantum efficiencies in excess of 30 percent, that is, 30 percent of the light which enters the telescope is recorded. This is contrasted with system averages like 0.01 percent 20 years ago. Thus, for many measurements, the 200-inch telescope is now 3000 times more efficient – and the advances in our understanding of the universe have grown to match. (The difference is dramatically illustrated in Figure 6.4, where a 5-minute CCD exposure is contrasted with a $1\frac{1}{2}$-hour photographic exposure on a very good photographic plate.)

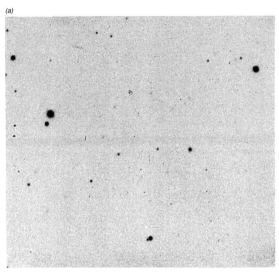

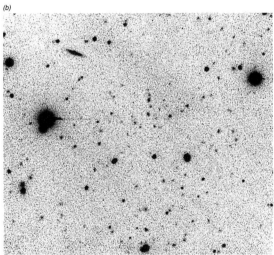

Figure 6.4 Two images of a cluster of galaxies at a redshift of 0.92, the light from which left it when the universe was about 3/8 its present age. To illustrate the power of CCD detectors: image (*a*) was taken on a very sensitive photographic plate with an exposure of $1\frac{1}{2}$ hours; image (*b*) to the same scale and of the same region, is a 5-minute exposure with a CCD. The cluster is the V-shaped group of images just above and to the right of the center. This is a negative print, in which stars and galaxies appear dark; more detail can be seen in such pictures than in normal ones. (Photographs from KPNO and Palomar Observatory.)

Today's large telescopes – most notably the Hale, the 4-meter telescopes at Kitt Peak, Cerro Tololo, Chile, and Siding Spring, Australia, the 3.5-meter ones at La Silla, Chile, and on Mauna Kea, and the Lick Observatory's 3-meter one on Mount Hamilton, California – have, through constant attention to detector development and instrument design, dramatically extended our view of the universe. Redshifts for bright normal galaxies can now be routinely recorded out to distances at which the universe was half as big and only 1/3 as old as it is now. Spectra in the faint outer parts of nearby galaxies have been successfully recorded to show that essentially all large galaxies and perhaps most small ones are surrounded by enormous halos of the mysterious dark matter. At large distances, this material contributes nearly all the mass, but, as far as we can tell, no light whatever. Quasars have been found at distances corresponding to redshifts in excess of 4, or, an epoch when the universe was only 1/5 its present size and 1/10 its present age. At that distance, the visible universe contains about 20 percent of the matter that we can ever see, given instruments of arbitrary power. (Most viable models of the universe have a so-called particle horizon, which corresponds to infinite redshifts but only finite distance, from which light traveling since the Big Bang has only now reached us. We can seen no farther: the particle horizon grows as the universe expands, but becomes twice as big only when the universe is twice its present age.)

We stand on a glorious precipice, able to see very far into the past, but not quite well enough to study objects in much detail. With little or no improvement expected in detector efficiency, the only choice is to build larger optical telescopes.

At wavelengths longer than optical, however, the detector revolution continues. In the infrared, at wavelengths to about 5 micrometers, about ten times that of visible light, imaging detectors are just becoming feasible. The first results are especially exciting for cosmological studies because at large redshifts galaxies emit the bulk of their energy in the near infrared. Hitherto there has been no way to study them there with any efficiency. At longer infrared wavelengths a terrible obstacle arises, because one is considering the same radiation emitted abundantly by objects at ordinary room temperature – like the telescope, the instrument, and, perhaps worst of all, the atmosphere. Studying faint stars in the daytime in ordinary visible light is, by comparison, easy. The only efficient way to circumvent this problem is to send into space an entire telescope cooled to a very low temperature. The recent spectacular success of the Infrared Astronomy Satellite (IRAS) mission, in which a very small telescope cooled to the temperature of liquid helium surveyed the whole sky in the wavelength range 12–100 micrometers, illustrated vividly the importance of going into space for infrared astronomy.

At yet longer wavelengths, we enter the realm of classical radio astronomy. At wavelengths of one to a few millimeters lie the spectral lines associated with the rotational motions of hundreds of common molecules. These molecules are found abundantly in dense, cold, dark clouds of gas and dust out of which stars (and presumably planets) are formed. Studying this radiation has told us much about the conditions and motions in these clouds, and giving tantalizing hints about how stars are born. From the beginning, the United States led this very technologically driven field; and, in the late seventies, a 25-meter antenna of very high accuracy was planned to study this region. It was never funded, and now the Japanese and Europeans have taken over the field. The Japanese have built an excellent large array of telescopes at Nobeyama, and the Europeans have a 30-meter dish in Spain, a high-accuracy 15-meter on Mauna Kea, and another 15-meter dish in Chile. Caltech has built a very accurate 10-meter telescope on Mauna Kea to study the almost entirely unexplored submillimeter wavelength region.

Unquestionably, the premier instrument for the study of the radio sky at wavelengths longer than a centimeter is the Very Large Array (VLA) on the plains of St. Augustine in New Mexico (Figure 6.5). It consists of 27 antennae, each 26 meters in diameter, in a Y-shaped configuration which can be extended (the telescopes move on special rails to change their positions) to a total distance from one tip of the Y to another of nearly 17 miles. In its special wavelength band, this telescope can produce pictures with higher resolution than those obtained by the largest optical telescopes. There is hardly an area of astronomy where the VLA has not had an important impact. Perhaps most fascinating is the revelation for the first time of the details of the double radio sources, in which the energetic particle beams from the central engine (black hole?) finally dump their enormous energies into the tenuous matter of intergalactic space (Figure 6.6).

Figure 6.5 An aerial view of the VLA radio telescope in its most compact configuration, showing most of its 27 individual antennae. The antennae move on railroad tracks and the configuration can be extended so that the tips of the Y are nearly 17 miles apart for the very highest-resolution observations. (NRAO/AUI photograph.)

Now under construction is the VLBA mentioned earlier, which will study relatively bright sources at unprecedented angular resolution. In doing so it may finally explore those regions near the big black holes that presumably power quasars and radio galaxies. There is some hope that the array can be combined with telescopes in Canada, Europe, and Russia to produce a truly globe-girdling network to reveal even finer detail. Preliminary experiments have met with qualified success; and, if the political problems can be overcome in such a way that the experiments can proceed on a regular basis, the results should be excellent indeed.

The major thrust in ground-based research for now seems to be toward large optical telescopes of entirely new design. The Hale telescope, as we have seen, was the model for a flock of smaller instruments, but its basic design is unsuited to instruments very much

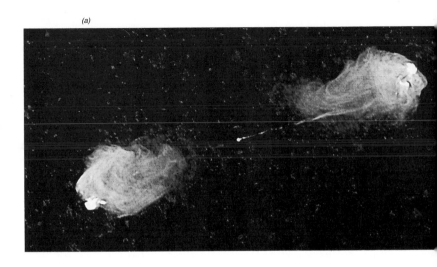

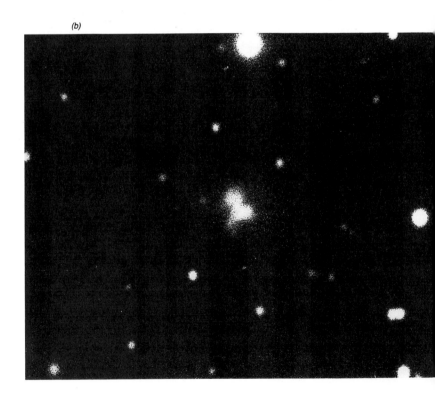

larger than itself. Yet, to study distant, faint galaxies, we need much larger telescopes. A rough rule of thumb is that for instruments of a given basic design, the cost is proportional to the 2.7 power of the aperture, nearly proportional to the weight. The Hale telescope would cost roughly $50 million today; one twice the size would cost $325 million, even if it could be built. (It couldn't!) Such sums seem out of the question for ground-based research, so a great deal of effort has gone into ways to make telescopes lighter. Since the telescope is basically just a framework to hold a mirror and a detector, the telescope can be made lighter if the mirror can be made lighter. Three approaches to the problem have emerged, and all the various projects now underway or on the drawing board use one of them.

The Hale telescope's Pyrex mirror was innovative at the time. Cast with deep ribs, it weighs only about 1/3 as much as a solid mirror of the same size. The first and most conservative approach, then, is simply to push the Hale design farther by making the ribbed sections of the mirror thinner yet. The overall shape of such a mirror follows classical design, but the mirror is 90 percent or more air. This design has the enormous additional advantage that the time required for the mirror to come to thermal equilibrium with its surroundings is very short, less than an hour. Since even a fraction of a degree temperature difference between the mirror and the air can cause disturbances at the mirror surface which seriously degrade the image quality, this is of great importance. The Hale mirror takes the better part of the day to reach equilibrium, which is still not too bad. By contrast, the great 6-meter Russian telescope, with its *solid* Pyrex mirror, realizes that winter has come about the first of April and that summer has arrived in early October. It is not too surprising that image quality has been a severe problem with that instrument – so severe that little useful science has emerged.

Figure 6.6 (*a*) A VLA image of the powerful double radio source Cygnus A. The very thin particle beams which power the giant lobes can be clearly seen. (NRAO/AUI radiograph.) (*b*) The optical image shows none of this activity, but only a somewhat peculiar galaxy – or perhaps two in collision. HST images may clarify what is happening. (Hale Observatories photograph.)

The lightweight honeycomb approach has been championed by Roger Angel at the University of Arizona, who has built a facility to cast such thin-rib mirrors up to 8 meters in diameter. The early ones will be used by a series of universities and foundations in various interlocking consortia to build telescopes in the United States and Chile. And, the National Optical Astronomy Observatories eventually hopes to build a multi-mirror telescope using four 8-meter mirrors for an equivalent aperture of 16 meters, a design patterned after the sextuple 4.5-meter Multiple Mirror Telescope of the Smithsonian Institution and the University of Arizona.

A rather more ambitious approach technologically is being pursued in the Keck Telescope on Mauna Kea, a project well underway by Caltech and the University of California (Figure 6.7). This 10-meter instrument uses not one mirror nor several independent ones, but many segments which fit together as a large reflecting mosaic. Each segment is individually supported and, with exquisite electronic control, maintains its position under changing gravity and wind conditions with respect to its neighbors. Such a design would have been unthinkable even a few years ago, so complex is the control system; but, advances in computers and mechanisms today make it almost straightforward – if still far from easy.

The European Southern Observatory is using yet another approach for its proposed Very Large Telescope (VLT). An array of four, independent 8-meter instruments, each telescope will use a single mirror, which is very thin and hence very flexible (Figure 6.8). To maintain its shape, the support for the mirror must be actively controlled, in much the same manner as the Keck telescope. In that case, however, it is easy to measure the position of any segment with respect to another and know exactly how to adjust its position. In the VLT that task is very much more difficult; and, in this author's opinion, this approach is the least desirable of the three and the least likely to succeed.

The new mirror technology is combined with new optical designs to reduce the length for a given aperture (the dome for the 10-meter Keck is about the same size as the dome for the 200-inch Hale telescope) and new ways of mounting the telescope. Classically, astronomical telescopes have been mounted with one axis parallel to the Earth's rotation axis (therefore pointed nearly at the North Star)

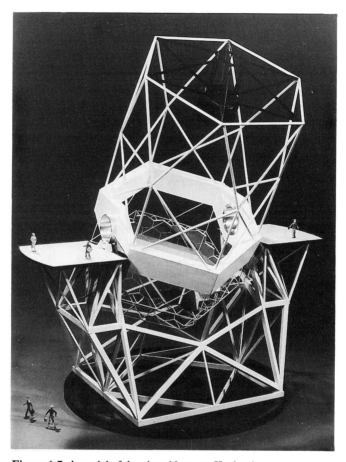

Figure 6.7 A model of the giant 10-meter Keck telescope now under construction on Mauna Kea, Hawaii. This instrument will use a segmented primary, the individual pieces of which are moved under computer control to behave like one large single mirror. (CARA photograph.)

and the other perpendicular to that. This has the advantage that, to track a star as it rises and sets, one need only rotate the instrument about the polar axis. By contrast, all the new designs use an altitude–azimuth mounting, with one axis vertical and the other horizontal. This greatly simplifies the loading due to gravity on the structure, but makes tracking very complicated, indeed, nearly impossible until a

Figure 6.8 A model of the VLT planned for the southern hemisphere by the European Southern Observatory. (ESO photograph.)

few years ago. However, computer-controlled tracking is now a simple task, even for a microcomputer. Always the push is to keep the weight down, to keep the structure simple. The Keck will cost about a $100 million – less than a 1/3 what a conventional telescope its size would cost. Other 8-meter Angel telescopes may be built for about $30–40 million.

Interestingly, all the large optical projects in the United States now underway or seriously being considered are mostly, if not entirely, funded with private money. The VLBA is being built with funds from the National Science Foundation, but has faced delay after delay and will probably not be finished before the middle of the next decade. Meanwhile, the major Japanese and European projects are entirely government-funded and are moving rapidly ahead.

Astronomy from space: present and future(?)

The space environment offers a large number of advantages for the pursuit of astronomical research, plus a few serious problems. First (and perhaps foremost) there are large ranges in the electromagnetic spectrum over which the Earth's atmosphere is completely or nearly opaque. All wavelengths shorter than visible light fall into this category, at least until one reaches gamma rays of very high energy. Thus, the whole ultraviolet, X-ray, and soft gamma-ray region is forever denied the ground-based observer. In the far-infrared, there are 'windows' of varying quality, but here one has the additional problem, discussed above, that the atmosphere *radiates* at the same wavelengths one is trying to observe. Even in the optical and near-infrared region, space provides a markedly darker sky. (Unfortunately, the sky from space is not as black as the popular notions would have it, being illuminated by sunlight scattering off interplanetary dust, but it is certainly much darker than the night sky seen from the Earth's surface.)

For imaging of any kind, the absence of an atmosphere has another enormous advantage. Without a medium in which to have thermal disturbances, the 1 arcsecond limit on angular resolution set by turbulent 'seeing' from the ground disappears. For sufficiently excellent optics, the resolution is set only by the size of the telescope and the wavelength of the radiation being studied. For the HST, 2.4 meters in aperture, the image size will be about 0.06 arcseconds, a factor of nearly 20 better than can be obtained routinely from the ground, and a factor of 10 better than the very best conditions at the very best sites.

The heaviest weight on the other end of the scale is cost. The HST will cost, by the time it finally flies, a billion and a half dollars, making it probably the most expensive single scientific project to date. (The proposed Superconducting Supercollider particle accelerator will cost much more, but the money has not been spent yet – most of the HST money has!) A great deal of that cost went into developments of general interest for the space program, but we still cannot 'charge off' *most* of the sum to that. The Keck telescope, recall, is expected to be built for about $100 million – a 1/15 the bill for HST, for a telescope with 17 times the collecting area. The reasons for the vast difference are many, some the outcome of good technical reasons, some very much the product of peculiar relationships between government and industry, some a straightforward product of the size of the overseeing bureaucracy. All are unavoidable within the framework of the current government-industry-academic structure in the United States. We will have more to say about this at the end.

There have, of course, been many spectacular successes in space astronomy already. A series of small X-ray satellites, beginning with Uhuru and culminating with the Einstein Observatory launched in 1978, have given us a window onto the realm of very high temperatures and energies. Sources as varied as clusters of galaxies, in which the tenuous gas between the galaxies has been heated to a temperature of a 100 million degrees, quasars, where very-high-energy particles are abundant, ordinary dwarf stars, where giant flares make high-temperature gas and high-energy particles, the neutron star and black hole companions of ordinary stars, where infalling material is heated to a 100 million or a billion degrees, all are copious emitters of X rays. The Einstein Observatory carried a camera which made pictures with resolution approaching that obtainable in the optical from the ground. Both the Einstein Observatory and its immediate predecessor HEAO-1 carried spectrographs which, for the first time, obtained information about the chemistry of X-ray emitting regions. An unfortunate decision at the beginning of the Einstein program to use gas jets for attitude control of the satellite meant that the instruments were still healthy when the steering gas ran out of power after 2 years of operation. No observatory of comparable power has been launched since; and, the only X-ray satellites active now are two small Japanese instruments.

In the ultraviolet, the International Ultraviolet Explorer (IUE) has been one of the most productive astronomical satellites ever launched. Now in its eleventh year and badly crippled by various small failures over its lifetime, it plugs away, producing spectra of stars, galaxies, and quasars in an octave-and-a-half ultraviolet region blocked by the Earth's atmosphere. In this region, stars hotter than the Sun radiate most of their energy, and the strongest absorption lines produced by abundant elements in the interstellar gas lie here, too. Thus, IUE has broadened our understanding of these areas immensely. Before IUE was the Copernicus Observatory, also a spectrographic instrument designed primarily for very-high-precision studies of the interstellar absorption-line problem using very bright stars. This role will be taken up by the HST when it is launched.

At wavelengths shorter than X rays, progress has been in fits and starts, with data obtained from some small satellite experiments, balloons, and rockets. However, there never has been a powerful, long-lived facility dedicated to astronomical research in gamma rays. One such instrument, called the Gamma-Ray Observatory (GRO), is now in advanced phases of construction.

Obviously, the premier instrument of the future, and the one scientists have been eagerly awaiting since planning first seriously got underway some 20 years ago, is Space Telescope. It began life as a somewhat larger and more ambitious instrument – the Large Space Telescope. It is now smaller and renamed the Hubble Space Telescope in honor of the many pioneering achievements of Edwin Hubble in extragalactic astronomy – or perhaps in some hope that it might finally measure the Hubble Constant.

HST is really the first serious attempt to build a long-lived space observatory, which, given the enormous expense involved, seems only appropriate. The telescope has been closely linked with the Shuttle program almost since its inception, a connection more politically than scientifically motivated and one that seems, especially now, of highly dubious wisdom. In any case, plans call not only for the Shuttle to launch HST, but for her astronaut crews to replace failed or obsolete instruments or other components, and to make necessary adjustments in the HST orbit. During the planned lifetime of 15 years, perhaps three generations of instruments will be installed. Currently, a launch is scheduled for late 1989, but that date is tied inextricably with the timetable for the resumption of Shuttle flights. Perhaps the safest thing to say now is that no one really knows when it will fly.

Many, many aspects of the HST have pushed the bounds of technical possibility. For example, to acquire pictures with 0.06 arcsecond resolution in exposures which may last for a large fraction of an hour requires pointing accuracy and stability many times better than ever achieved before. (For reference, 0.06 arcsecond is the size a penny appears 21 miles away.) The image of a star at the focus of the telescope is about 6/10 000 of an inch in diameter, and the instrument recording that image, which weighs several hundred pounds and is typically the size of a telephone booth, must not move or creep during the exposure even a small fraction of that amount. As a result of these requirements, exotic materials and exotic techniques abound in the satellite.

HST is also a large and heavy satellite, at the limit in both size and weight of what the Shuttle can place in orbit; it was these limits which caused the reduction in size from the original 3-meter to the current 2.4-meter (94-inch) aperture. Its length with the protective cover closed is 43 feet, the diameter of the aft shroud containing the scientific instruments 14 feet, and the total weight 12.5 tons – about a 1/3 that of a typical ground-based telescope of the same size (Figure 6.9). Most of the optical assembly is made of graphite epoxy, famous

Figure 6.9 An artist's conception of the HST in orbit. (NASA illustration.)

in 'high-tech' tennis racquets for its resilience. In HST, the epoxy is used for its high strength, light weight, and, most important, its ability to be made with a thermal expansion coefficient of zero, so the mechanical assembly will not change size as it heats or cools (Figure 6.10). The mirror is a fused quartz honeycomb, similar in principle to the Pyrex honeycomb mirrors of Roger Angel, but made with a better quality (and more difficult-to-handle) material. The mirror's surface is finished accurate to a 1/50 of a wavelength of visible light, or about 1/500 000 of an inch. The HST mirror is probably the largest optical element of this accuracy every made. (The US Air Force also has some very good large mirrors designed for looking down rather than up, but their sizes and specifications we can only guess.)

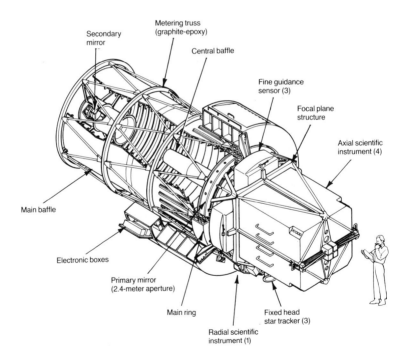

Figure 6.10 A cutaway view of the HST, showing the telescope assembly, the scientific instruments, and the fine guidance sensors. Most of the structure in this illustration is made of graphite epoxy, titanium, and invar, a steel which has a very low thermal expansion coefficient. (NASA illustration.)

The complement of instruments to be launched with the telescope include two spectrographs, two cameras, and a photometer. The spectrographs are primarily aimed at doing ultraviolet spectroscopy, though one of them, the Faint Object Spectrograph (FOS), can also work in the visible. The spectrographs basically extend the work done by the IUE satellite to higher accuracy and much fainter objects. The High Resolution Spectrograph (HRS) will again be used predominantly for studying the interstellar medium via absorption lines, but with much higher resolution than employed by Copernicus and IUE. The FOS will be able to take ultraviolet spectra of even very faint quasars. For distant quasars, we can see the ultraviolet part of their spectra because it is redshifted into the visible band. By contrast, very little is known about the ultraviolet spectra of nearby quasars because their redshifts are so small the light is still in the ultraviolet and hence inaccessible from the ground. Moreover, all but the very brightest are too faint to study with the IUE satellite. We therefore do not know whether these nearby objects are at all like their far-away cousins. HST will unravel that mystery and, perhaps more importantly, will allow us to study the intergalactic medium nearby using quasar light as a probe. Absorption spectra from hydrogen and other elements like carbon are seen in abundance in the spectra of high-redshift quasars. This suggests that clouds containing some of the elements synthesized in stellar thermonuclear processes existed in intergalactic space early in the history of the universe. We do not know whether these clouds are connected with galaxies which have already formed, are in the process of formation, or have nothing to do with galaxies at all. The phenomenon cannot be studied nearby because the lines are in the ultraviolet. With HST, we will be able to see the lines in nearby quasars, if they exist, and will be able to study directly the galaxies (if any) with which the lines are associated.

In my mind, the most exciting instruments on HST are the cameras, at least partly because I have been closely associated with one of them for the last 10 years. The vistas that will be opened to use with HST's improved resolution are staggering.

The two cameras are: The Faint Object Camera (FOC), an instrument being constructed as part of an international agreement with the European Space Agency, which is also supplying the solar power panels on HST; and, the Wide-Field/Planetary Camera (WF/PC), for which I serve as Deputy Principal Investigator. The goals of the two instruments are complementary. The FOC has a very narrow field of view and is designed to make observations of the very highest spatial resolution; the WF/PC has slightly lower resolution but offers a field which will include the whole of moderately distant galaxies or very distant clusters of galaxies, and offers CCD efficiency. (The FOC uses a more conventional television sensor with much lower quantum efficiency.)

Pictures from HST will tell us:

> *What very distant galaxies look like.* The pictures in Figures 6.4 and 6.11 of distant galaxy clusters taken from the ground show the galaxies as barely discernible smudges, even in the

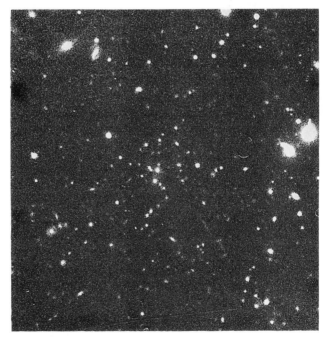

Figure 6.11 A CCD image of a cluster at a redshift of 0.756, similar to the one shown in Figure 6.4(*b*), taken under very good conditions with the Hale telescope. The galaxies show almost no detail, with the structure lost in the blurring caused by turbulence in our atmosphere. (Palomar Observatory photograph.)

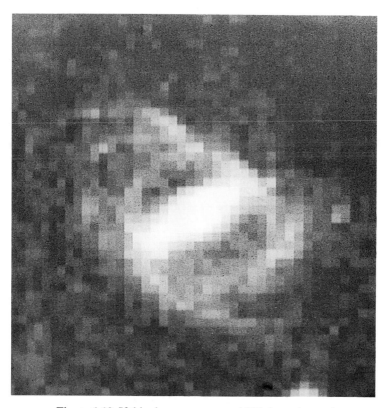

Figure 6.12 If this cluster were moved 100 times closer, the galaxies, one of which might look like this, could be studied with ease. But, to study them in their youth, and hence compare them with similar objects nearby to see how they evolve, means that we must study them at great distances. (James Gunn photograph.)

CCD frame. Figure 6.12 shows what one might look like in a cluster at the same redshift, with a long HST exposure. Such pictures can be compared with images of nearby objects to look directly for the evolution of galaxy forms predicted by many theories.

What kind of galaxies are hosts to the quasar phenomenon. From the ground, the glare of the bright quasar in the nucleus of the parent galaxy makes it well-nigh impossible to study the galaxy itself. With HST's order-of-magnitude improvement in resolution, the galaxy will be an easy target.

We do not know whether quasars form in normal galaxies, or only in very unusual objects. We also do not know the current source of fuel for the quasars. HST pictures should provide answers to both questions.

What kind of stars there are in nearby galaxies. Figure 6.2 shows the outer regions of the Andromeda Spiral, essentially the only big galaxy other than our own in which we can study individual stars. With HST, pictures of this quality can be obtained for many galaxies ten or more times distant. Perhaps these photographs will allow us finally to get accurate distance measurements to enough galaxies to solve the distance scale problem.

How material accumulates to form a star. In the nearest star-forming region, some 100 light years away in the constellation of Taurus, one can, by taking data over the lifetime of HST, study in detail the motion of material around very young stars. The interpretation of such data will not be easy because of the shrouding dust, but the possibility of actually watching stars form and perhaps learning something about how the process occurs is immensely exciting.

Whether nearby stars have planets. As Jupiter circles the Sun, the Sun, a thousand times more massive, moves slightly in response. As a result, the Sun follows a slightly wavy path through space corresponding to the period of Jupiter's orbit, about 12 years. Any similar wobble in the paths of nearby stars is too small to detect from the ground; but, with the exquisite images of HST, it should just be possible. For the first time, we may know whether solar systems like our own are common, as most astronomers believe, or, in fact, are very rare.

The HST is only the first in an ambitious series of proposed satellites which NASA has dubbed 'The Great Observatories.' If the 20-odd-year lead time from serious study to launch for HST is predictive, and if costs continue to spiral upward so we cannot even contemplate working on more than one project at a time, we will all be dead before much of this comes to fruition. However, let us at least think about the possibilities. The range covers essentially the entire electromagnetic spectrum, from the gamma-ray region (GRO) to millimeter radio waves (the Large Deployable Reflector), with some consideration of a space-borne centimeter-wave antenna to complement the ground-based VLBA

The GRO is the most highly developed of these missions and also the cheapest; it may actually fly in the early 1990s. It will study the energy range from about 100 000 electron volts to 20 billion electron volts, a much larger range than previous experiments on the small COS-B and SAS-2 satellites, and with about a factor of 20 higher sensitivity. Very little is known about cosmic sources of gamma rays, and GRO can confidently be expected to provide us with many surprises. While GRO will address a few definite problems, its primary mission is to explore a new and very interesting realm of the most energetic processes in nature. Such exploration has never failed to be rewarding, and we can expect nothing less in this case.

Three other instruments rival HST in complexity and, almost certainly, cost. While their potential for discovery is extraordinary, their chances of coming to fruition are a little doubtful at this time. The first, and the highest priority on the list of projects outlined for the 1980s by the National Academy of Sciences, is an X-ray observatory called AXAF, the Advanced X-Ray Astrophysics Facility. This imaging instrument is similar in concept to the earlier Einstein satellite, but now designed like HST for a long life and with instruments which can be changed and updated. Its higher optical quality, larger effective aperture, and more sensitive detectors will make it about a factor of 100 more sensitive than Einstein and capable of producing images of 1/2 arcsecond or better. AXAF will answer many questions, including the origin and fate of the hot gas which seems to pervade most clusters and groups of galaxies in the universe. Understanding this seemingly technical point is crucial to knowing how clusters and galaxies originally formed and discovering how we came to be. It should be able to detect essentially all of the quasars in the observable universe, thus giving us vital information about how and when they formed and what role they play in the formation and evolution of ordinary galaxies.

Just as the AXAF takes over from HST at the short-wavelength end of the spectrum, an instrument called SIRTF will extend observations in the longer wavelength infrared regime. SIRTF, an acronym whose meaning has conveniently evolved over the years and now stands for Space Infrared Telescope Facility, is a cooled telescope of about 1-meter aperture. With its 1000-fold increase in sensitivity, this observatory will expand upon work begun with IRAS. It will be able to study galaxies at very large redshifts, perhaps even detecting primeval galaxies in their initial burst of star-forming activity. (These objects have eluded detection from the ground for the last 30 years, either because they are very faint or do not exist – the significance of either case to understanding how galaxies form is very great indeed.) SIRTF will also be able to study star-forming regions with great sensitivity.

Finally, in the far infrared and submillimeter radio-wave region is an ambitious instrument called LDR, the Large Deployable Reflector, a 10-meter-diameter antenna which will be erected in orbit, it being too large to fit into the Shuttle bay. Uncooled, except perhaps passively with a sunshade, LDR will allow high-resolution imaging to wavelengths as short as 10 micrometers. It will be ideally suited to studying the star-forming problem in much greater spectral and spatial detail than can be done from the ground or any other proposed satellite. It is much less well developed, as well as technologically more difficult than either AXAF or SIRTF, but it promises much, including experience in assembling instruments in orbit. We would eventually like something like a 10-meter *optical* telescope in space, and LDR could pave the way.

Summary and prospects

Wild optimism about the future of observational astronomy seems well warranted now. Plans are everywhere for telescopes larger than any before. At least two such projects are substantially funded and one is well underway. Despite some formidable technical problems still to be faced, there is little doubt they will succeed. Essentially perfect detectors exist for the optical region, and surely it will not be long before they are available in quantity and optimized for the kind of very low-light-level applications astronomers need. The HST is essentially finished and is simply awaiting the resumption of Shuttle flights. The combination of giant spectroscopic instruments on the ground and the superb images from HST promises to revolutionize our understanding of galaxies, the distance scale in the universe, and any number of other perplexing problems now facing astrophysics.

I think the above is a not unreasonable view, but a little caution is in order. Since the Challenger disaster, NASA has (quite reasonably) become extremely safety-conscious. The HST is very heavy and was to have been launched with higher-than-nominal thrust from the main Shuttle engines. (In any case, HST can only be launched with Atlantis, the lightest Shuttle.) Whether this will be allowed to happen now is not clear. The launch has been delayed so long that the solar activity cycle will be at a radically different phase than first planned – near maximum instead of near minimum. The combination of lower thrust and the increased solar activity means both that the orbit will not be as high as planned and that the atmosphere of the Earth will be more extended. The first orbit will therefore be short-lived, and the eroding effect on the optics could well be very severe. Indeed, if it is to survive, HST will need to be boosted to higher orbit with another Shuttle flight, probably within the first year. To have to count on *two* Shuttle flights in the post-Challenger atmosphere of almost paranoid caution is to me singularly frightening.

The outlook for projects in other wavelength regions seems completely grim. On the ground we are doing well in the optical band only because of the generosity of private donors. The one, large, ground-based radio project actually being built is the VLBA – and it is being so starved for funds that it may well never be finished. The United States has completely fumbled the ball with millimeter radio astronomy, and is fortunate that others in the world have recognized its importance.

The whole US space program is in shambles. We (the nation and NASA – not, I may say with some bitterness, the scientific community) foolishly chose to tie that program to the small Shuttle fleet. When one failed, with the unfortunate loss of the lives of all the crew, the US space program effectively ceased to exist for some period yet to be determined. The lesson – that one must have some sort of alternative launch capability – seems not to have been learned. Indeed, even the maliciously false myth that Shuttle launches would be cheaper than rockets seems not to have died in the face of contrary fact. The disaster froze an already timid bureaucracy, and the effect has been a total loss of nerve.

Perhaps that loss of nerve was coming, even without Challenger, egged on by ever-worsening costs and delays. Sadly, America's civilian space program is supplied by the same vendors who supply the high-tech military programs – and that sector is concerned little with cost and seemingly even less with delay. Obviously, the firms involved found it both attractive and profitable to treat all their customers on the same footing.

Science, to be sure, is not blameless. In the early days of space exploration, experiments were simple and often constructed by the researchers themselves in university laboratories; therefore, they were also often cheap. This approach has become less and less possible as experiments have become more sophisticated. (Of course, the enormous complexity of some experiments is in part a response to the 'It's gotten so expensive that we can only have one flight for X years; we have to make it do *everything*!' Syndrome, which is obviously self-fulfilling.) The reliance on a manned Shuttle has also put severe constraints on simple, cheap experiments. The documentation and safety and accountability requirements is such that most university laboratories and small firms simply cannot cope with the mountains of paper. Another part of the problem is psychological: Nearly all of us do science because of a burning need to know how the world is made and functions. We are used to having to be clever, having to use ingenious solutions, and having to face difficult engineering problems. Nature responds, sometimes brutally, sometimes surprisingly kindly, to our probings. But to reach her only through a ponderous bureaucracy, or a commercial complex whose goals and motivation have little to do with ours, is deadly. Only the worst of us win the battle.

Happily, things seem more cheerful elsewhere. The Europeans have had their problems with Ariane, but, at least, a single Ariane failure does not paralyze the entire program. Other European programs have led to a sophisticated X-ray observatory called ROSAT and an infrared satellite called ISO, both of which will fly and gather data while the more powerful and ambitious AXAF and SIRTF remain gleams in their promoters' eyes. The Japanese are still in the era of cheap and simple experiments, which is one reason American investigators look upon them with such envy. Meanwhile, they have the only X-ray satellites flying at all. It is worth remembering that X-ray astronomy can *only* be done from space. The last American satellite died 7 years ago, and AXAF, even with the best of efforts, will not fly before the late nineties. The field will not survive, and the skills will die.

The Russians, although pursuing strong and vigorous space programs, seem to be doing little astronomy. However, they seem poised for major advances in the difficult areas of measuring cosmic microwave background and the far infrared. Perhaps they will jolt the United States to life.

In any case we must do something, and do it soon, or space science in the United States will die. Since many of the problems faced by US scientists are experienced by researchers in other countries, they might be solved simply by real international cooperation at a very low level in the political–industrial hierarchy. More likely, it will require some thoroughgoing change in the way modern science is done. How it will be resolved I cannot know or even guess. Nothing very practical has suggested itself as yet. I am not, I fear, very optimistic.

Further Reading

Cornell, J. and Carr, J. *Infinite Vistas: New Tools for Astronomy*, New York: Scribner's, 1985.

Pioneering the Space Frontier: The Report of the National Commission on Space, New York: Bantam, 1986.

CONTRIBUTORS

James Cornell is Publications Manager of the Harvard–Smithsonian Center for Astrophysics. He is the author or editor of several books on astronomy, including *The First Stargazers*. Currently, he is president of the International Science Writers Association.

Margaret Geller is an astrophysicist at the Smithsonian Astrophysical Observatory and Professor of Astronomy at Harvard University. Her research interests include extragalactic astronomy and cosmology.

James Gunn is the Eugene Higgins Professor of Astrophysics at Princeton University. In addition to studies in theoretical astrophysics, Gunn has made observations with large ground-based telescopes, and has helped develop astronomical instrumentation, including detectors for the Space Telescope.

Alan Guth is a Professor of Physics at the Massachusetts Institute of Technology and a physicist at the Smithsonian Astrophysical Observatory. His primary research interest is the application of particle physics to studies of the very early universe.

Robert P. Kirshner is Professor of Astronomy at Harvard University. His research interests include the study of supernovae and supernovae remnants. Previously, he was director of the McGraw-Hill Observatory.

Alan P. Lightman is Professor of Science and Writing at the Massachusetts Institute of Technology and a physicist at the Harvard-Smithsonian Center for Astrophysics. His research interests include compact objects, radiative processes, and stellar dynamics. He is also the author of several books for general readers, including *A Modern Day Yankee in a Connecticut Court*.

Vera C. Rubin is a member of the Department of Terrestrial Magnetism of the Carnegie Institution of Washington. Her research interests are in galactic and extragalactic astronomy. As the member of several advisory panels and councils, Rubin has played an important role in shaping American research policies and objectives.

Subject index

Advanced X-ray Astrophysics Facility
(AXAF), 178, 181, 182
age of the universe, 152
estimate of, 49
Andromeda Galaxy (M31), 2, 3(illust.),
23, 82, 82(illust.), 93, 150,
151(illust.)
motion towards Milky Way, 149
angular resolution, 149, 157
apparent brightness, 30
variation with distance, 30
application of physical laws, 84
arcsecond defined, 27
Aristotelian universe, 4, 5
aether, 4
motionless Earth, 4
Ptolemy's modification of, 5
Assyrian cylinder, 1, 2(illust.)
axion (particle), 71

Babylonian gods, 1
Marduk, 1
Ishtar, 1
Ti'amat, 1
Babylonian maps, 50, 51(illust.)
background radiation, 99
microwave, 111
baryon number, 130–2, 135
baryonic matter, 102
Bell Telephone Laboratories, 111
Big Bang, 2, 25, 46, 70, 71, 74, 97, 98,
100, 106, 108, 110–13, 115,
117–19, 130, 135, 138, 151, 152,
156, 158, 161, 162
big crunch, 118
birth of universe, 2
black hole, 156, 158, 162

as quasar power source, 157
maxi, 101, 104
mini, 101, 104
Bootes, 65
Bootes Void, 47, 47(illust.)
brown dwarf stars, 101
bubbles, 63, 116
bulge stars, 75, 78, 84

catalog of galaxy positions, 57
Cepheid variable stars, 22, 23, 30, 34, 46
use in distance measuring, 32, 33
in LMC, 31
in Virgo Cluster, 33
Centaurus A, 79(illust.)
supernova in, 79
Cerro Tololo Interamerican
Observatory, 88, 161
chains, 101
charge coupled devices (CCD), 44, 90,
158, 159
clocks, 13
synchronism, 13
clusters, 83
distribution, 83
formation, 83
fragmentation, 83
Coma Cluster, 58, 61, 94, 102
continental drift, 53, 54, 72
continent fitting, 53
Copernican system, 7, 8, 9
Kepler's acceptance of, 8
Copernicus observatory, 171
core collapse, 78
cosmic background radiation, 112, 116,
117
directionality, 117

homogeneity, 116
inhomogeneities, 117
ratio of photons to protons and
neutrons, 116
spectrum, 112(illust.)
temperature measurements, 116
cosmic expansion, 134
cosmological constant, 107, 134, 139
footnote
critical density, 66, 98, 118

dark halo, 93
dark matter, 62, 66, 67, 71, 73, 74,
93–7, 99, 101, 104, 140, 155
gravitational effects, 73
missing mass, 73
deceleration parameter (q_0), 73
decoupling, 70
density perturbations, 119
difficulty of measuring motion, 87
disk, 84
disk stars, 75
distance measurement, 15, 29, 35
difficulties, 35
early estimates, 15
distribution of galaxies, 58(illust.)
Doppler shift, 22, 41, 43, 87, 107, 148
blueshift, 87
galaxy rotation, 43, 87
line-of-sight velocity, 87
redshift, 87

early universe constituents, 113
Earth, 5, 6, 13, 74, 169
cosmos centered on, 13
effects of atmosphere on observations,
169
radius measurement, 5, 6
Einstein observatory, 170
Einstein's static universe, 18–21, 106
electromagnetic field, 128
electromagnetic spectrum, 169
electromagnetism, 120, 121, 129
interaction, 120, 121, 129
elementary particle physics, 119, 120
use by cosmologists, 119
elements, 40, 78, 115
abundancies, 115
characteristic colors, 40
production, 78
same across the universe, 40
elliptical galaxies, 95
equation for orbital velocity, 92
European Southern Observatory, 166

escape velocity, 153
expanding universe, 2, 16, 21–4, 40, 59,
62, 97, 111, 138, 148
discovery of, 40
early speculations, 16
redshifted background radiation, 111

false vacuum, 134, 135, 137
fiber optic spectrograph, 46
filaments, 101
future evolution of the universe, 98

galaxies, 42, 67, 70, 75, 78, 82, 84, 86, 91
brightness distribution, 86
clusters, 67
distribution of, 70, 84
Local Group, 82
mass determination, 86
mass distribution, 86
rotation curves, 89, 91
spectra, 42(illust.)
spirals, 75, 78
bulge stars, 75
disk stars, 75
major components, 75
star formation in disk, 78
star orbits, 75
Gamma Ray Observatory (GRO), 171,
178
gamma rays, 158
general relativity, 106, 107, 121, 122,
135, 136, 138
description of gravity, 106, 121
Grand Unified Theories (GUTs), 105,
110, 119, 122–5, 127, 129, 131–4,
141–5
gravitational
energy, 136
instability of the universe, 119
interaction, 120
repulsion, 134, 135
gravitationally bound universe, 66
gravity
falling apple and, 85
Galileo's experiment, 85
Gum Nebula, 77(illust.), 78
star formation in, 78

Hale telescope, 154, 158, 159(illust.),
161, 163, 165, 166
Harvard College Observatory, 31
studies of LMC, 31
helium, 41
Hereford map, 50, 52(illust.)

hierarchy problem, 124
Higgs field, 127–9, 134, 137, 143
    mechanism, 127
High Energy Astrophysical Observatory
    (HEAO-1), 170
homogeneous expansion, 108,
    109(illust.)
honeycomb mirrors, 166
Hubble age, 152
Hubble Constant (H_0), 44, 60, 73, 97,
    98, 107, 108, 110, 149, 150, 152,
    153, 171
    variability of, 107, 152
Hubble flow, 102
Hubble's Law, 44, 46, 48, 107, 110,
    111, 115, 148, 151
Hubble Space Telescope (HST), 29, 33,
    149, 150, 152, 154, 155, 169, 171,
    172(illust.), 172–80
    instruments, 174, 175
    launch and servicing, 171, 180
    observations, 175–7
    size, 172
    structure, 173
Hubble velocity, 152

inflation, 66, 100, 132, 133, 135–43,
    145, 150, 153
Infrared Astronomical Satellite (IRAS),
    162
infrared imaging detector, 161
Infrared Space Observatory (ISO), 181
interactions of nature (forces), 120
International Ultraviolet Explorer
    (IUE), 170
inverse square law, 86
isotropy of background radiation, 97

Jupiter's moons, 14
    as clocks, 14

Keck telescope, 46, 154, 166–9
Kepler's Laws, 147
Kitt Peak National Observatory, 88, 161

La Silla, 161
Large Deployable Reflector (LDR),
    178, 179
Large Magellanic Cloud, 31, 80,
    81(illust.), 103
    cepheids in, 31
    distance to, 31
    supernova 1987A, 31, 80, 103
Large Space Telescope, 171

Las Campanas Observatory, 88
light year defined, 2
Local Group, 81, 82
    as member of Virgo Supercluster, 82
    number of members, 81
    types of galaxies in, 31
Lowell Observatory, 22
low-luminosity galaxies, 88

M31, 32
M33, 32, 33(illust.)
M87, 95, 96
    mass, 95
Magellanic Clouds, 80, 81
    encounters with Milky Way, 81
    fragmentation of, 81
    future of, 81
    hydrogen tidal tail, 80
    passage through Milky Way disk, 80
magnetic monopoles, 133
mass, 62, 85, 90, 92, 93, 95, 110, 118
    determination from orbital velocity, 92
    distribution in galaxies, 90, 93
    distribution in solar system, 85
    missing, 62, 95
    universe, 110, 118
        uncertainty, 110
mass–luminosity ratio, 95, 96
Mauna Kea, 46, 154, 161, 162
Mercury, 85
microwave background radiation, 70,
    71, 102, 139
    photon scattering, 71
Milky Way, 2, 31, 74, 75, 76(illust.), 80,
    149
    components of, 75
    dust in, 75
    galaxies beyond, 75
    Galileo's observations, 75
    satellite galaxies, 80
    shape of, 74
motions of stars, 85

nebulae discovered, 13
negative pressure, 137
neutral hydrogen, 93
    21-cm line observations, 93
neutrinos, 102, 103
    as dark matter, 102
neutron star, 156
Newtonian gravitation, 62, 85, 90, 101,
    106, 136, 138
    mechanics, 107

NGC 801, 88(illust.)
  spectrum of, 88(illust.)
NGC 891, 75, 76(illust.)
  compared to Milky Way, 75
NGC 7293, 79(illust.)
NGC 7331, 150(illust.)
Nobeyama array, 162
nucleosynthesis, 98, 99, 111, 115, 118

observational cosmology, 73, 74
  developments in, 74

Palomar Mountain, 154, 158
Palomar Sky Survey, 58, 155
parallax, 13, 27–30, 49
  angle, 27
  measurement limits, 28, 29
  61 Cygni, 27
particle physics, 105
Perseus–Pisces region, 65
Planck scale, 143, 144
planetary nebula, 78
Pluto, 85
primordial fireball, 97
Principia, 15
protostar, 78

quartum chromodynamics (QCD), 120, 121
quarks, 120
quasars, 157, 158, 161, 163
  black hole as power source, 157

radio astronomy, 162
radio galaxies, 163
rate of universal expansion, 113
rates of nuclear reactions, 113
redshift, 22, 46, 49, 59, 60, 155, 161, 179
  measurements, 60
relic radiation, 73, 97
  discovery of, 73, 97
  motion relative to, 97
ROSAT, 181

Sanduleak −69 202, 35, 37
sheets, 63, 66, 68, 83
Siding Spring, 161
Small Magellanic Cloud, 31
small scale structure, 119
solar system, 13, 78
  formation, 78
  measurement, 13
Soviet 6-m telescope, 165
special relativity theory, 106

speckle interferometry, 37
spectroscope, 41
spectrum, 41
speed of light, 117
spontaneous symmetry breaking, 122, 125, 133, 137
standard candles, 30, 49
star clusters, 30
  distance measuring, 30
star evolution, 78
stereo vision, 27
strings, 101, 144
strong interaction, 120, 121, 129
Sun, 74, 85, 156
  mass determination, 86
  non-uniqueness, 28
superclusters, 84
supercooling, 133
supernovae, 35, 37, 78
  core collapse, 36
  distance indicators, 35
  shell expansion, 35
  Type I, 35
  Type II, 37
  1987A, 35, 37
superstrings, 144
$SU(2)$, 121, 122, 127–9, 133
$SU(3)$, 121, 122, 127–9, 133
$SU(5)$, 122, 142

telescope, 10, 166, 167
  first used for astronomy, 10
  mountings, 166, 167
theoretical cosmology, 74
threads, 63

Uhuru, 170
unified electroweak theory, 121
universal expansion, 107
unstable nuclei, 113
$U(1)$, 121, 127–9, 133

Very Large Array (VLA), 162, 163(illust.), 168, 180
Very Large Telescope (VLT), 166
Very Long Baseline Array (VLBA), 157, 163, 178
Virgo Cluster, 83(illust.), 95, 96, 157
voids, 61, 63, 65–8, 83, 101, 103, 116
  dark matter in, 68
volume of visible universe, 54

white dwarf stars, 101
wimps, 71

X-ray,
  galaxies, 96
  gas, 95
  halo, 96
  observations, 95

satellites, 170, 178, 181, 182
  AXAF, 178, 181, 182
  Einstein, 170
  ROSAT, 181
  Uhuru, 170

Name index

Albrecht, Andreas, 132
Angel, Roger, 173
Aristotle, 4, 5, 9, 10, 25
Arp, Halton, 96
Audouze, Jean, 99

Baade, Walter, 108
Bardeen, James M., 140
Berkenstein, Jacob, 101
Bentley, Richard, 16
Bertola, Francesco, 96
Bessel, Freidrich Wilhelm, 27
Brahe, Tycho, 37, 80, 147
Bruno, Giordano, 10

Cassini, Jean-Dominique, 13, 14, 27
Copernicus, Nicholas, 7, 8, 9
Cowie, Lennox, 68, 102
Cowsik, Ramanath, 103
Cronin, James W., 132

da Costa, Luis, 65
de Lapparent, Valerie, 60, 83, 102
Descartes, Rene, 10, 15, 16
de Sitter, Wilhelm, 19
de Vaucouleurs, Gerard, 96
Dicke, Robert, H., 119
Digges, Thomas, 10

Eddington, Sir Arthur S., 23
Einasto, Jaan, 63
Einstein, Albert, 17, 18, 19, 20, 21, 25,
    106, 107, 134
Eratosthenes, 5, 10

Fabricant, Daniel, 95
Fitch, Val L., 132
Flinders, Matthew, 53

Forman, William, 96
Forman-Jones, Christine, 96
Friedmann, Alexander, 20, 21

Galilei, Galileo, 10, 13, 75, 85, 147
Geller, Margaret, 16, 46, 83, 102, 116,
    155
Georgi, Howard, 122
Givanelli, Riccardo, 65
Glashow, Sheldon, 121, 122
Gorenstein, Paul, 95
Gunn, James, 29, 107
Guth, Alan, 16, 66, 99

Harrison, Edward R., 141
Hawking, Stephen W., 140
Haynes, Martha, 65
Herschel, William, 16
Higgs, Peter W., 127, 134, 137
Hooke, Robert, 85
Hubble, Edwin P., 21, 23, 24, 32, 44,
    59, 107, 108, 152, 153, 171
Huchra, John, 44, 59, 60, 83, 102
Humason, Milton, 44

Ikeuchi, Satora, 68
Ishtar, 2

Jarvis, John, 70

Kant, Immanual, 16, 17
Kepler, Johannes, 8, 37, 80, 147
Kirshner, Robert, 65, 102, 107
Koo, David, 70

Lambert, Johann, 16
Laplace, Pierre Simon, 17
Larson, Richard, 96

Leavitt, Henrietta S., 22, 32
LeMaitre, Georges, 23
Linde, Andrei D., 132
Lorentz, H. A., 18

Magellan, Ferdinand, 31
Marduk, 1, 2
McClelland, John, 103
Milgrom, Mordehai, 101
Millikan, Robert, 20
Minkowski, H., 18
Muir, Jon, 80

Newton, Isaac, 10, 14, 15, 16, 17, 19,
    23, 85, 106, 138, 147
Nisenson, Peter, 40

Oemler, Augustus, 46, 102
Oort, Jan, 94
Ostriker, Jeremiah, 68, 102

Papaliolios, Costas, 40
Peebles, P. J. E., 119
Penzias, Arno A., 70, 111
Pi, So-Young, 140
Ptolemy, 5, 8, 52

Quinn, Helen R., 122

Rankine, William, 19
Roberts, 93, 94(illust.)
Rubin, Vera, 43, 54, 62, 67, 94(illust.),
    154

Sakharov, Andrei D., 130
Salam, Abdus, 121
Sandage, Allan, 108
Sanduleak, Nicholas, 35, 37
Schechter, Paul, 46, 102
Shectman, Stephen, 46, 102
Schramm, David, 99
Shapley, Harlow, 23
Slipher, Vesto Melvin, 22, 44
Starobinsky, A. A., 140
Steinhardt, Paul J., 132, 140
Szaley, Alex, 70

Thomson, William (Lord Kelvin), 20
Ti'amat, 1, 2
Tinsley, Beatrice, 96, 99
Turner, Michael S., 140, 145
Tyson, Tony, 70

Uson, Juan, 71

Valdes, Frank, 70
van den Bergh, Sidney, 36

Wegener, Alfred, 53
Weinberg, Stephen, 121, 122, 130
Weyl, H., 18
Wilkinson, David, 71
Wilson, Robert W., 70, 111
Wright, Thomas, 16

Yoshimura, Motohiko, 130

Zel'dovich, Yaakov, 62, 141